Book 4

Team Success Network - Enhancing Mutual Success Teams Using AI Tools

Use the New AI Advantages Never Before Available

By Mr. Elijah J Stone
and the Team Success Network

Team Success Network - Enhancing Mutual Success Teams Using AI Tools

Team Success Network Publishing
Email: connect@teamsuccessnetwork.com

Published by Team Success Network Publishing
Printed in the United States of America

ISBN (Paperback): 978-1-968654-04-7
Library of Congress Control Number: 2025913824
First Edition: June 2025

Cover Design by: Team Success Creative
Interior Layout by: Team Success Creative

Contact Us – Let's Build Together

This is more than a book. It's an invitation. A call to rise, to act, to build—together.

The "Team Success Network" exists to support churches, ministries, and Christian groups as they pursue God's vision: bringing Heaven to Earth through unified, purpose-driven action. If this message stirs something in you, or if you're ready to explore what mutual success looks like in your community—we want to hear from you.

Let's start the conversation.

Reach out here:

📩 General Connection – connect@teamsuccessnetwork.com

📩 Share Your Team's Results & Testimonies – results@teamsuccessnetwork.com

📩 Submit Documents or Project Materials – docs@teamsuccessnetwork.com

📩 Contact Elijah Personally – elijah@teamsuccessnetwork.com

We're here to support, encourage, and collaborate. Whether you're just starting out or leading a growing movement—this is your first step. Let's build something eternal.

Table of Contents

PART 7: People – The "Mutual Success Team" All Using AI

PART 8: Relationships – Unity, Support – The Health of the Team

PART 9: Making An Impact

Part 1: The Foundation – With AI – For "Team Success" – To Our Christian Advantage

This foundational section introduces the relationship between AI and Christian living, showing how believers can leverage this new tool in faith. It challenges the fear narrative and instead offers a spiritually grounded confidence in using AI under God's guidance. The goal here is not to replace our reliance on the Spirit—but to multiply our obedience with power and precision.

Readers will discover that AI, when used wisely and prayerfully, acts like a supportive assistant—not a leader. When each team member learns how to use it as a tool for planning, organizing, and accelerating progress, the result is clarity and collaboration. Every person in the "Mutual Success Team" becomes more efficient and focused, freeing up time for true Kingdom work.

It sets the stage for a new mindset: one where believers not only reject fear but step forward boldly to innovate within their faith-led projects. This includes how AI can serve in daily spiritual life and in larger team strategies. When used with faith, AI becomes a vessel for multiplying impact—not an obstacle to be feared.

This part lays a spiritual and practical foundation that empowers churches, teams, and individuals to begin exploring and growing with AI. It reminds us: God equips us with every good tool at the right time—and today, AI is one of them.

CHAPTER 1 – Team Success + AI: Unlocking New Levels of Power for Individuals and Churches

Multiply What's Possible in Your Life, and Across the "Mutual Success Team" Network

---

Personal Power: Using AI to Strengthen Your "Team Success"

It's an incredible time to be alive as a believer. We have never had so many resources, so much information, and such immediate access to tools that help us follow through on what God puts in our hearts. And now, with the rise of AI, we've been handed something truly unique: a free, fast-thinking assistant that is always ready to help.

Imagine this—you're called to write a letter to encourage someone, but you don't know how to begin. You open a free AI tool, explain what you're trying to say, and within seconds, you're staring at a loving, well-crafted draft that reflects the very heart of what you wanted to communicate. That's the kind of assistance now available to anyone. Not just pastors or tech-savvy leaders—but to every believer, every day.

It helps with more than just communication. AI can help you brainstorm new business ideas. It can take your handwritten notes and organize them into a clean action plan. It can help summarize Scripture passages, find Bible chapters related to a topic, or even create a morning routine that keeps you spiritually strong and practically focused. With AI, it's not about doing more work—it's about working with more clarity and confidence.

And as your personal life becomes more ordered, your contribution to your "Mutual Success Team" also begins to shift. You start showing up to meetings with ideas already outlined. You offer solutions backed by strategy. You create shared documents that benefit the whole team. This is what it looks like when every believer becomes empowered—not just to know more, but to build more. That's the power of personal transformation multiplied by technology, fueled by faith.

You also begin to experience something even more powerful—consistency. AI helps you follow up on goals you've set, habits you want to strengthen, prayers you want to track, and projects you need to finish. It becomes your daily companion for forward motion. When AI helps with structure, you gain space to breathe. When it helps you execute your ideas, you gain momentum. And when it helps you encourage others, your influence expands.

Over time, this consistent personal momentum begins shaping not only what you accomplish, but who you are becoming. You step into each day better prepared. Your mind becomes clearer. Your priorities feel aligned. You begin carrying yourself with more peace and less stress. All of this flows outward, empowering those around you—including your family, your team, and your community. And that's how transformation starts, one empowered believer at a time.

Equipping the Church: Using AI to Accelerate Coordination and Outreach Across "Mutual Success Teams"

Now imagine that same level of support flowing into your church or "Mutual Success Team." The weight of ministry often falls on a few shoulders—but AI can lift that weight and help distribute it across the entire team. It doesn't replace leadership. It amplifies what good leaders are already trying to do.

One of the biggest challenges in "Team Success" coordination is simply staying on track. Who was supposed to follow up with the new visitor? Where are the notes from the last training? Has anyone prepared flyers for the upcoming outreach event? These aren't spiritual problems—they're logistical ones. And they are exactly the type of problems AI is built to solve.

With just a few simple tools, your team can automate reminders, generate content for newsletters and social media, track who's been contacted and who still needs care, and summarize team meetings into a list of next steps. Everything becomes easier to manage. What used to take 10 hours a week might now take 2. And the time you save? That goes back into relationships, prayer, teaching, and outreach.

AI can even help your ministry scale globally. It can translate sermons and training documents into 100+ languages. It can summarize testimonies in formats that are easy to share. It can align calendars across time zones for international calls and prayer chains. Suddenly, your local team has global reach—and none of it sacrifices your local voice. Why? Because you train AI to speak your language. You feed it your mission, your values, your examples. And then it simply carries that same voice further and faster than you ever could on your own.

Think about the impact: Churches that once struggled to document their breakthroughs can now preserve every win. Ministries that couldn't afford translation teams can now serve multilingual communities. "Mutual Success Teams" that used to spend hours each week coordinating now flow with ease and focus. This is what happens when you pair spiritual clarity with digital tools.

AI also helps unify vision across departments. Worship leaders, youth teams, volunteers, and pastoral staff can all stay updated with the same shared documents and dashboards. Nothing slips through the cracks. Everyone moves together. This kind of clarity strengthens morale, improves communication, and reduces unnecessary duplication of effort. It creates a stronger, more joyful ministry experience for everyone involved.

When AI helps establish a common foundation, everyone can work in alignment—faster, lighter, clearer. Ministry teams experience fewer dropped tasks, more fulfilled commitments, and stronger trust among leaders. The same people doing the same things now feel energized instead of drained. That is what better coordination produces—momentum, harmony, and capacity to do more for the Kingdom.

The Breakthrough Principle: You Don't Have to Do Everything—You Just Have to Start

This isn't about becoming tech experts. It's about becoming faithful stewards of the resources God has already placed in our hands. The goal isn't to replace ministry—it's to free it. Your starting point can be as simple as using AI to write a follow-up message, draft a meeting summary, or organize your goals for the month.

Pick one task—just one—and ask: *Would this be easier with help?* Then give it a try. As you begin to experience what these tools can do, your confidence will grow. You'll see how to use them for both personal growth and collective mission. You'll even begin sharing tips with your team, showing others what's working—and that is how transformation begins to ripple across the whole "Team Success Network."

What once felt overwhelming now becomes manageable. What once seemed impossible—creating a coordinated network of believers across towns, cities, and nations—becomes real. And it's not because of the technology alone. It's because of what happens when technology is guided by faith, purpose, and the wisdom of the Holy Spirit.

When you gain support for the areas that normally slow you down, you accelerate in every area that matters. You gain clarity, speed, and focus. You lead more effectively. You serve with less strain. And your ability to show up for others multiplies. That's what this chapter is all about—finding ways to do what God has already called you to do, with greater strength, joy, and endurance.

The more you use AI as a support tool, the more your bandwidth increases. You'll go from managing to multiplying. From staying afloat to building strong. From thinking small to dreaming bigger. And when your work becomes smoother, your heart has more room to listen to God, love people, and stretch toward your highest purpose. That's what we're here to do.

Let God Stay at the Center—and Let AI Support the Work

No one is saying that AI replaces the Holy Spirit. In fact, it's the opposite. The more we inquire of the Lord, the more clearly we see where and how to use the tools He's made available to us. The Spirit still leads. God still gives vision. But when He opens doors, He often places tools in front of us—so that we can walk through those doors faster and stronger.

AI can help eliminate wasteful hours. It can help preserve testimonies. It can help structure training. It can even help you identify people in need, follow up with compassion, and steward the growth God is sending your way. This is not hype—it's help. And we must not be afraid to receive it.

You are part of something big. You're not only building your own life—you're building alongside a global family of believers committed to mutual success. These tools help you do that better. They help your "Mutual Success Team" move faster, care deeper, plan clearer, and share stronger. And as we work together—with the Spirit of God leading and the tools of the age supporting—we will see what Jesus declared come to pass: *"The Kingdom of Heaven is near."*

Let's bring it even closer—together.

Chapter 2 – Peaceful AI: Let Fear Go—and Embrace the Leverage—of a New Team Partner in Your Mutual Success Team

Trusting God, Letting Go of Fear, and Working With the Tools in Our Hands

---

Letting Go of Fear in the Face of Change

Fear has a way of sounding wise when it's really just stalling progress. It asks, "What if this backfires?" or "Is this dangerous?" But as believers, we are not called to live in fear—we are called to walk by faith. Especially when something is new or unfamiliar. Artificial Intelligence (AI) may seem strange or even threatening to some. But when used rightly, it doesn't replace God's power—it supports the people who carry His presence.

We must remember 2 Timothy 1:7: *"For God has not given us a spirit of fear, but of power and of love and of a sound mind."* This applies not just to inner emotions, but to decisions. If we're filled with the Spirit of God, then we're also filled with the capacity to discern, evaluate, and wisely use every good tool God places in our hands—including technology.

Fear keeps us from taking new ground. It slows down church systems, stunts growth, and limits our vision. That's not faith—that's hesitation dressed up as discernment. Don't let fear block a breakthrough. Instead, consider that AI might be one of the very tools God is giving to accelerate what He's already called you to do.

AI Has Limits—And That's What Makes It Safe

Let's set the record straight: AI isn't alive. It can't pray. It can't love. It doesn't hear from God. It doesn't care about people. And because of that—it *can't replace you*. The fear that AI will "take over" Christian ministry is based more on sci-fi than scripture. What AI actually does is process information. It repeats patterns. It suggests ideas. It follows your input.

There are important safeguards in place. Developers have built-in filters, oversight tools, content monitoring, and ethical standards. Governments are stepping in too, requiring transparency and accountability. And on top of all that, as believers, we carry something no machine ever will: the Holy Spirit. Prayer, wise counsel, and biblical values provide a higher filter that AI can never mimic.

You don't need to fear AI because you don't let AI lead. You lead—with faith, scripture, and the Spirit. AI just helps carry the weight. It assists. It writes. It organizes. But it does not direct. That's your job, and God is still speaking. As long as you're listening to Him first, you're already ahead of the curve.

A Tool to Fulfill Vision—Not Replace the Mission

What if AI could give you 10 extra hours a week? What if it could write your notes, schedule your meetings, summarize your outreach, and translate your flyers—so you could spend more time discipling, praying, or resting?

That's not a fantasy. That's already happening for hundreds of churches, ministries, and Mutual Success Teams around the world. And not because they trust technology more—but because they trust God *with* their technology.

In Exodus 4:2, God asked Moses, *"What's that in your hand?"* Moses had a staff—and God used it. You have a smartphone, a laptop, and an internet connection. Could those be the staff in your hand today?

AI isn't magic. It won't build the vision for you. But once God gives you the vision, AI can help execute it with speed, clarity, and excellence. Flyers get written in seconds. Meeting notes get summarized instantly. Testimonies get translated into 15 languages. That's not replacement—that's Kingdom leverage.

Inquire of the Lord First—Then Move in Peace

Here's the key: pray first. Don't rush. Don't fear. Just ask the Lord, *"Should we use this?"* If yes, move forward boldly. If not yet, wait with peace. When you start by inquiring of the Lord, you set yourself up for clarity, not confusion.

Proverbs 3:6 tells us, *"In all your ways acknowledge Him, and He will direct your paths."* That includes tech tools. You're not being reckless for exploring AI—you're being faithful for asking first. And once God gives peace, use that peace as permission to move.

AI doesn't replace intimacy with God. It doesn't replace human obedience. It doesn't replace prayer. But once you've heard God clearly, it can help you walk out that assignment faster and with greater reach. This is not rebellion—it's stewardship.

So ask God to replace fear with wisdom. Ask Him to make you sensitive to what He's saying. Then allow Him to show you how to use AI as a faithful, Spirit-aligned partner for multiplying your Mutual Success Team's effectiveness.

AI Helps You Multiply Testimonies and Teachings

One of the biggest gaps in the Church isn't anointing—it's documentation. We have powerful moments... but we don't always remember how they happened. We teach great things... but don't always share the material. AI can help us capture what God is doing—then scale it to others.

Let's say your team just held an outreach that saw 25 salvations. AI can help write a recap, turn it into a newsletter, generate social posts, and format it into a PDF so others can copy your strategy. In minutes, you've turned one win into a model for dozens of churches.

The early Church grew through testimonies. The book of Acts is essentially a collection of Spirit-led stories. And each one encouraged, trained, and empowered others. That same principle works today—when we're wise enough to record what God is doing.

AI helps your team turn every success into a system, every story into a tool, every event into a blueprint. When paired with the Spirit, this isn't just progress. It's multiplication.

Final Word: Let Fear Go and Let God Use the Tools

Fear will always whisper "what if." Faith responds with "even if." Even if AI changes the world, God is still in charge. Even if systems advance, Jesus still saves. Even if technology evolves, the Kingdom remains unshaken.

You don't need to fear what God can redeem. In fact, sometimes the enemy's greatest tactic is to keep believers afraid of the very tools that would give them an advantage. But we are not of those who shrink back. We are those who build.

So use AI—not as a crutch, but as a multiplier. Use it to lighten the load. Use it to reach further. Use it to train faster, document better, and testify louder.

The Kingdom of Heaven is near. The mission hasn't changed. But the tools have advanced—and God is ready for His Church to lead with wisdom, excellence, and boldness.

You are not behind. You are not alone. You are fully equipped. And now, with peaceful AI in your hand, you are ready to build what Heaven is revealing.

Chapter 3 – AI-Powered Mutual Success Teams: Automating What Slows the Church Down

There's nothing more frustrating than a God-given vision getting stuck or slowed down.

The truth is, most churches and Christian groups already know what needs to be done. The problem isn't vision — it's time, energy, and organization. That's where AI comes in. Not to take over, but to **remove the bottlenecks.** That's what this chapter is about.

Mutual Success Teams are meant to move fast, share resources, and solve needs. But sometimes even the most willing group gets buried under scheduling, writing, or endless paperwork. AI can lift that burden.

When the busywork is handled, the Kingdom moves faster.

What Slows Churches Down the Most

Most delays in church growth and outreach aren't spiritual. They're logistical. You've seen this happen:

- Planning meetings get rescheduled three times.
- No one has time to write a solid project proposal.

- The one person who knows how to make a flyer is already burned out.
- A great business idea sits idle because no one made the checklist.

These are not attacks from the enemy. These are fixable gaps in time, tools, and focus. And they slow down real Kingdom impact.

Tip: Don't let divine assignments get buried under human limitations.

What AI Can Do (Starting Today)

With one or two free AI tools, you can automate a huge part of what's been slowing you down:

- Need a **grant proposal** written? AI can do a first draft in 3 minutes.
- Want to **schedule** 10 meetings? AI can suggest the plan and generate follow-ups.
- Need a **business flyer, FAQ sheet, or onboarding document**? AI can create those based on your prompt.
- Want to **document** the steps of your Mutual Success Team project for others to duplicate? AI can organize and write that for you.

All you need is a clear idea. The AI fills in the structure, saves time, and gives your team a major head start.

Tip: Think of AI as your digital assistant. It's not in charge. It's just ready to serve.

Automating the "Repeatable Stuff"

The real power of Team Success comes when your model is repeatable.

But that means someone has to write it down, organize it, and explain it. That takes time. Unless you let AI do it.

Use AI to create:

- **Step-by-step instructions** for others to copy your project
- **Standard email templates** for outreach or recruiting help
- **Schedules** and job roles for project volunteers or interns
- **Social media posts** to promote your success story

When one Mutual Success Team finishes a great project, AI can instantly help you export that as a guide for others — even in another country.

Tip: Automate what's already working, so others can repeat it faster.

Multiply Results, Not Burnout

When churches try to do everything manually, growth is slow and stressful. But when they use AI to handle admin and organization, it frees people to pray, lead, serve, and build relationships.

That's what we actually want from church anyway.

Let the Holy Spirit move. Let people minister. Let visionaries dream. Let AI do the formatting and organizing.

Tip: Don't pour out your people doing tasks that tech can now handle.

Final Word: Free Up the Flow

Mutual Success Teams are built to accelerate what the Church is already called to do. AI just removes the friction.

When you remove delays, you unlock flow.

That means more people served. More needs met. More cities reached. More leaders trained. More Kingdom impact — at less cost, and in less time.

That's how you redeem technology.

Not by replacing people. But by freeing them.

Tip: Use AI to remove what slows you down — so God can speed up what He's building through you.

Chapter 4 – Prompting With Purpose: Hearing God and Acting in Faith Within Your Mutual Success Team

Understanding Prompts, Discernment, and Spirit-Led Conversations With AI

What Is a Prompt? And Why Does It Matter to Christians?

You may not have heard the term before, but when you use AI—especially tools like ChatGPT—you interact with it through something called a **prompt**. A *prompt* is simply the question or instruction you give the AI. It's the way you start a conversation with it. It's how you shape what comes back.

If you type, "Help me write a newsletter," that's a prompt. If you say, "Give me a list of scriptures on healing," that's a prompt. If you ask, "Create a checklist for launching a Bible study with three churches," that's a prompt. In simple terms, a prompt is *what you say to start the process.*

And here's the beautiful connection for believers: God also responds to prompts. We prompt heaven when we pray. We prompt the Spirit when we ask, "Lord, what should I do?" And in this season, we're learning to prompt *with purpose*—in both the natural (AI tools) and the supernatural (our walk with God).

Understanding how to prompt wisely can help you access insights, clarity, strategy, and speed. When you align your prompting with God's voice, you're not just being efficient—you're being Spirit-led.

From Casual Prompts to Strategic Requests

There's a big difference between throwing out a random question and asking something with clarity and intention. That's true in prayer—and it's true in AI.

When you prompt an AI like ChatGPT, Claude, or Gemini, it responds based on what you've asked. The better your prompt, the better the response. If you say, "Give me five ideas for outreach," that's fine. But if you say, "Give me five outreach ideas that a Mutual Success Team of three churches in Kenya could do in May, with no budget,"—you'll get far better, more relevant, more usable answers.

This isn't about sounding smart. It's about being specific. And it mirrors what Jesus taught when He said, *"Ask, and it will be given to you; seek, and you will find"* (Matthew 7:7). Asking is a kind of prompting. And asking with faith, focus, and obedience opens doors—in both realms.

AI is not divine. But it reflects a Kingdom principle: You have not because you ask not. So what will you prompt today? Will you just dabble—or will you strategically request something that moves your mission forward?

Let God Lead the Prompt

Before you start typing into any AI system, pause. Ask the Holy Spirit, "What do You want me to work on right now?" Sometimes we jump into action, trying to get help from AI, before we've even gotten help from God.

Here's a better rhythm:

1. **Pause** – Take 10 seconds. Breathe.
2. **Pray** – Ask, "Lord, what should I work on today?"
3. **Prompt** – Only *then* go to the AI with your next step.

This keeps you aligned. It keeps you from wasting time. It transforms AI from a noisy distraction into a useful assistant—serving the direction God already gave you.

In the same way a GPS works best when you already know your destination, AI works best when the Holy Spirit has already spoken the goal. AI helps with the *how*. But God still gives the *what* and *why*.

Team Success Prompting—How Groups Use AI Together

When you're part of a Mutual Success Team, AI can become a shared tool that each member uses daily. But to maximize it, you need a prompting strategy that includes:

- Shared language: Agree on how to describe your vision and projects to AI.

- Pre-set goals: Know what you're trying to build, automate, or explain.
- Assigned roles: Let each team member take ownership of prompting for different tasks (content, flyers, training, data insights).

Here's an example of a team-wide prompting session:

- One person prompts: "Write a 3-step guide for how our Mutual Success Team can host a healing outreach event for 50 people."
- Another prompts: "Create an invitation flyer for that same event."
- A third prompts: "Draft follow-up questions to ask attendees for testimonies and next steps."

Now you have a team that's operating in sync—with AI assisting all parts of the workflow. And it all began with simple, clear, purposeful prompts.

You Can Train AI to Align With Your Voice and Values

One of the most powerful prompting techniques is to *teach the AI who you are*. This builds consistency and saves time. You can start by writing:

"I lead a Christian Mutual Success Team focused on business and ministry. We believe in abundance, healing, generosity, and team success. Please keep your responses faith-filled, aligned with biblical principles, and in a tone that's warm and uplifting."

Once you set that up, every prompt you give after that will sound more like *you*. Whether you're writing newsletters, social posts, course materials, or business plans—you'll see your voice reflected. And over time, your results will only get better.

This is powerful. But it's also practical. And best of all—it's scalable. You can share these prompts with your whole team. Everyone can start getting high-quality results... today.

Examples of Prompts That Multiply Progress

Sometimes it helps to see examples. Here are a few prompts that Christian Mutual Success Teams are already using:

- "Summarize this sermon into 5 email devotionals for team members."
- "Give me a 7-day training outline for new volunteers joining our outreach team."
- "Create a flyer for a business-and-Bible fellowship happening next Saturday."
- "Write a follow-up message to thank attendees and encourage testimonies."
- "Give me ideas for a weekly rhythm that includes prayer, outreach, and business planning."

Notice the pattern? Each prompt is practical, Spirit-aligned, and built for forward motion. You can use them too. And once you see the impact, you'll want to prompt with purpose every day.

What Happens When You Prompt in Faith

This chapter isn't just about better questions. It's about better positioning. When you prompt in faith—believing that God will guide both your spiritual and practical tools—you position yourself to move faster and with greater clarity.

Hebrews 11:6 says that *"Without faith it is impossible to please God."* That doesn't just apply to salvation or healing. It applies to leadership. Strategy. Innovation. Prompting with purpose means prompting with faith.

It means saying, "Lord, I believe You've called me to this project. Now help me use every tool You've made available to carry it out."

When you start prompting this way, your Mutual Success Team becomes unstoppable—not because AI is powerful, but because your faith is.

Final Word: Let Every Prompt Be Spirit-Led

You don't need to know everything. You just need to know the One who does. The Spirit of God within you can guide your mind, your words, and yes—even your prompts.

Ask before you type. Clarify your goal. Aim your question. Expect wisdom. And then let your obedience to God shape every tool you touch.

You were made to lead with clarity. To prompt with purpose. And to move in faith, not fear.

Let this be your new habit: Spirit first. Prompt second. Then build.

Chapter 5 – Inquiring of the Lord: Hearing God Individually & Together in Mutual Success Teams

How the Holy Spirit Guides Us to Make the Right Decisions in Business and Life

---

God Is Well Able—So Ask Him First

In life and in business, the single most powerful habit you can develop is this: **ask God first**. Not after you're confused. Not once you've made the wrong move. But before you make the decision, before you launch the project, before you choose the direction—*inquire of the Lord*.

This isn't just good advice. It's a scriptural pattern. All throughout the Bible, the people who sought God's counsel in advance were the ones who experienced victory, clarity, and peace. The ones who didn't often ended up in unnecessary trouble.

You may have heard it said: *"God is well able."* That's not just a statement of faith—it's a strategy. If God truly knows all, and if we truly have access to His Spirit, then why wouldn't we pause and ask for His input before we act? Especially when we're leading a Mutual Success Team where our choices affect others.

When we inquire of the Lord, we activate a divine advantage. And when we do it *together*—as a team—the results are even more powerful.

The Unfair Advantage of Spirit-Led Decision-Making

Let's be honest: success in business often comes down to one thing—**making the right decisions consistently**. And here's the truth most of the world doesn't realize: **believers have an unfair advantage**. Not because we're better, but because we have something better.

We have the mind of Christ.
We have the wisdom of God.
We have the Spirit of truth living inside us.

So why do we hesitate to ask for divine guidance? Why do we rely only on logic or analytics when we have access to Heaven's insight?

Here's what we've learned:

- Most business failures don't come from bad effort—they come from bad decisions.
- Most good efforts still require God's timing and insight to produce good fruit.
- Many paths *look* right—but end in frustration, delay, or financial loss.

But when we inquire of the Lord, He shows us the better way. Sometimes that means waiting. Sometimes it means pivoting. Sometimes it means partnering with someone we hadn't considered. But every time, it leads to a win-win. Because **God's way always has a good outcome.**

Inquiring Together: Mutual Wisdom in Mutual Success Teams

When you're working in a team, the stakes are even higher. What you decide affects not just you—but your church, your partners, your business allies, and the people you serve. That's why **corporate inquiring**—seeking God together—is so essential to Mutual Success Teams.

Here's how you can do it:

- Begin each major planning session with 5 minutes of shared prayer
- Pause during meetings and say, "Let's ask the Lord about this"
- Give room for silence and spiritual impressions
- Take note of any scriptures, confirmations, or peace that arises
- Don't rush consensus—seek clarity

When a Mutual Success Team stops to inquire, God responds. And that response brings alignment. Suddenly, what seemed confusing becomes clear. What seemed risky becomes resolved. God's guidance doesn't just help—it *unifies* the entire team.

You're not just working together. You're *hearing* together. And that's where supernatural strategy begins.

The Role of the Baptism in the Holy Spirit

Many believers today love God, read the Word, and even pray regularly—but they're not operating in the full power available to them. The Baptism in the Holy Spirit is a gift from God that empowers us to **hear more clearly**, **pray more boldly**, and **serve more effectively**.

It's not just a doctrinal idea. It's a practical advantage in business and life.

When you're filled with the Holy Spirit:\n- You gain the ability to pray in tongues—which builds up your spirit and clears your mind

- You experience a heightened sensitivity to God's voice and direction
- You operate with bolder faith, deeper insight, and greater spiritual endurance

Here's a simple prayer you can pray, if you've never received this gift:

"Father, I thank You that the baptism in the Holy Spirit is for me. I receive it now, by faith, in Jesus' Name. I believe I am filled with Your Spirit, and I expect to speak in other tongues and walk in Your power."

When you operate in that power daily—especially in business and team leadership—you stop guessing. You start discerning. And that's when true acceleration begins.

Why So Many Fail Without Asking

Let's get real: Many failures in Christian business and ministry happen because someone skipped this step. The opportunity looked great. The plan sounded reasonable. The door seemed open. But they didn't ask God.

That's how we get detoured, delayed, and discouraged.

In our experience, the biggest wins we've ever had were not because we were the smartest. They happened because we **inquired of the Lord**, and He showed us *the way to go*. That's not theory. That's our testimony.

If you skip this step, even good things can go wrong. But if you stop and ask—even when it feels inconvenient—you invite the God of the universe into your planning session.

What happens next? Miracles. Wisdom. Provision. Strategy. Favor. Peace. All because you asked.

What Joint Inquiring Looks Like in Real Life

You don't need to turn every meeting into a revival night. But you *can* make joint inquiring part of your team culture.

Here's what that might look like in a Mutual Success Team meeting:

- Someone says, "Let's take 3 minutes to ask the Lord what we're not seeing yet."
- The group pauses. Some pray quietly. Some journal. Some listen.

- Afterward, people share what they sensed: scriptures, pictures, cautions, or confirmations.
- The team checks for unity. If there's peace, they proceed. If not, they pause.

That's it. Nothing weird. Nothing spooky. Just *Spirit-led teamwork*. And over time, that habit becomes the most valuable practice in your entire organization.

Final Word: The Key to Life Is Hearing from God

We believe the key to success in life is simple: **always make the right decision**. And the only way to always make the right decision is to always ask the One who knows the right path.

You don't have to be perfect. You don't have to have all the answers. But if you develop the habit of *inquiring of the Lord*—both individually and as a team—you'll find yourself walking into favor, clarity, and results that only Heaven could produce.

And the world will wonder how you did it.

So here's the invitation: make inquiring your default posture. Let it be your first instinct, not your last resort. Invite God into every meeting, every project, every launch, every partnership.

He's already ready to speak.

Ask. Listen. Act.

And watch your Mutual Success Team move into the success God has already prepared.

Chapter 6 – When the Spirit Leads the Prompt: Using AI as a Faith-Filled "Mutual Success Team" Member in God's Plan

Allowing the Holy Spirit to Direct Your Questions, Your Workflows, and Your Wins—For Personal and Team Breakthrough

---

Introducing AI as a "Team Member" in Your Daily Kingdom Work

Imagine if you had a team member who never needed sleep, could work instantly, and had access to more information than a library of books. Now imagine that team member took your instructions seriously, responded in seconds, and never complained about the work. That's what AI can be—**not a leader, but a support member in your Mutual Success Team**.

We're not talking about replacing people. We're talking about expanding capacity. We're talking about **letting the Spirit of God lead the conversation**, even when it starts with a keyboard. Because when the Spirit leads *the prompt*, what comes back from AI can be incredibly useful, strategic, and Kingdom-focused.

What's a prompt? It's simply the question or command you give to the AI. But when the Holy Spirit inspires what you ask, and how you ask it, the results aren't just smart—they're *supernaturally helpful*. That's the sweet spot: AI directed by Spirit-filled believers, for the good of the Church and the growth of the mission.

AI Helps You in Two Dimensions—Personal and Team

You don't have to choose between your own success and your team's success. In fact, one will supercharge the other. **When you use AI to build your personal clarity, habits, and confidence**, you'll have more to contribute to your Mutual Success Team. And when you use AI to support your team's projects, planning, and communication, you create an ecosystem where everyone thrives.

Let's break this down:

For Personal Success

- Ask AI to help you structure your day for maximum productivity
- Prompt it to give you Scriptures and prayer plans for your current season
- Use it to summarize sermons or books you're trying to learn from
- Let it help you write vision documents or goal sheets with clarity and motivation

For Team Success

- Have AI create flyers, announcements, or outlines for your group events

- Generate training materials based on your team's specific projects
- Organize roles, timelines, and job descriptions with AI assistance
- Translate your content to reach broader audiences across languages

This isn't about being high-tech. It's about being Spirit-led and *strategically equipped*. The more you prompt, the more you build. The more you build, the more your team grows. The more your team grows, the more the Kingdom expands.

When the Holy Spirit Shapes the Prompt, the Results Multiply

The power is not just in asking AI questions. The power is in **what questions you ask—and why**. When you allow the Holy Spirit to influence your prompts, you begin creating from a place of wisdom, purpose, and alignment with Heaven.

Instead of typing, "Give me 10 outreach ideas," you pause and ask, "Holy Spirit, what do You want us to accomplish this month?" Then you prompt:

"Give me 10 outreach ideas that align with a heart for families, low cost, and high spiritual impact."

That's Spirit-led prompting. It starts with discernment. And it invites God into your creative process.

Another example: You might sense in prayer that your Mutual Success Team is called to raise up new trainers. Don't just guess what to do. Ask:

"Create a 3-week training module for new Christian leaders to help them launch a local Mutual Success Team, including vision casting, project setup, and accountability."

See the difference? The prompt isn't just technical—it's prophetic. You're building something in step with God's voice. And when you do that consistently, AI becomes more than a tool. It becomes **a digital helper that helps manifest what Heaven is already planning.**

AI Reflects Your Voice, But Heaven Gives the Assignment

Let's be clear: AI can't tell you what your calling is. It doesn't give you purpose. It doesn't carry anointing. **Only the Holy Spirit can give you divine assignment and clarity of vision.** But once you know what you're called to do—AI can help you structure it, scale it, and spread it.

That's the balance we're after:

- **Heaven reveals the "why"**
- **You and your team hear the "what"**
- **AI helps with the "how" and "when"**

You stay in the lead. The Spirit stays at the center. AI stays in its lane—as a helper, not a head.

This isn't just a smart strategy—it's a deeply spiritual one. It's the same as how Paul used letters to spread the gospel, or how modern churches use livestreams to reach millions. The *method* changes, but the *message* stays sacred.

So go ahead—train your AI assistant to reflect your language, values, and style. But let the Holy Spirit decide what the work is about in the first place. That's how we build with power and precision.

Examples of Spirit-Led Prompts for You and Your Team

To bring this to life, here are some practical prompts that Spirit-filled believers are already using—both for personal and team success:

For Personal Use:

- "Help me create a 21-day devotional plan on trusting God through uncertainty."
- "Give me a time-blocked schedule that balances prayer, work, and family."
- "List 7 questions I should reflect on this week to grow in my spiritual leadership."
- "Write a prayer for our team's breakthrough, based on Ephesians 3:20."

For Team Use:

- "Generate a flyer text for our upcoming community outreach in 3 different formats."
- "Create a 5-slide presentation for a Mutual Success Team training workshop."
- "Build a checklist for launching a group project with 3 churches and 5 volunteers."
- "Translate our project summary into Spanish, French, and Swahili."

As always: check everything. Refine it. Tweak it. Add Scripture. Let your spirit confirm. AI drafts, but *you* direct. The Spirit edits. You approve. Then you deploy.

Final Word: You're Not Replacing People—You're Extending Purpose

This chapter is not about idolizing technology. It's about **multiplying impact through wisdom and obedience**. When the Spirit leads your prompt, your team gets stronger. When your team uses AI faithfully, your personal life gets lighter. And when both grow together, the Kingdom gets a massive boost.

You're not asking AI to replace the people you love. You're asking it to **support the people you serve**. And when you do that with discernment, humility, and faith—you become a builder who moves fast, moves right, and moves in the Spirit.

Let your next prompt begin with prayer. Then give it purpose. Let AI help you move the vision forward—not just as a digital assistant, but as a *faith-filled tool in God's plan*.

He's building something beautiful through you. Don't build it alone.

Part 2: The Vision – "Mutual Success Team" Potential

This part casts a wide vision for how "Mutual Success Teams" can unlock a powerful future by embracing AI. It invites churches and ministries to imagine a world where no believer is left behind, and every gift is brought to the table. AI becomes a bridge, connecting believers with the wisdom and resources they need to succeed.

It encourages teams to stop limiting themselves to their current resources and start leveraging tools that multiply capacity. Through the lens of faith, AI becomes a clear partner—not in spiritual discernment—but in streamlining execution and clarifying goals. The aim is to move faster, with greater confidence, in fulfilling Kingdom assignments.

This part invites every reader to imagine what it would be like if entire church networks, families, and ministries operated with synchronized clarity and Spirit-led momentum. AI doesn't replace that vision—it helps bring it to life. The "Mutual Success Team" model is expanded into a scalable, repeatable format that inspires action.

As this section concludes, the reader understands that this isn't a gimmick or short-term tool. It's part of a global spiritual movement, where believers step into their full potential by pairing ancient truths with modern strategies.

Chapter 7 – Use All 4 AIs Together - ChatGPT, Grok, Claude AI, & Google Gemini

4 Experts Instead of Just 1 — Get an Unfair Advantage in your "Team Success"

If one AI is helpful, imagine what happens when you use four.

That's the mindset behind this chapter. Think of each free AI platform as its own unique expert—ready to help you answer questions, develop plans, write applications, solve problems, or brainstorm ideas.

You don't need a team of human assistants. You can build your own *Mutual Success AI Team*—and it fits in your pocket.

How It Works

You ask one question.

Then you paste that same question into four different apps:

- ChatGPT (**https://chatgpt.com**)
- Grok (**https://grok.com**)

- Claude AI (**https://claude.ai**)
- Google Gemini (**https://gemini.google.com**)

Each AI will give you a different answer—often in tone, detail, or strategy.

Now you have four different expert-level opinions. You compare them. You combine the best insights. You're more prepared than anyone else who only asked *one* source.

Tip: *Give yourself the gift of a broader perspective. It's free, fast, and ridiculously effective.*

Why 4 Is Better Than 1

Sometimes only *one* of the four will actually give you the answer you need.

That's why this method works.

If you rely only on one app, you risk hitting a wall. But if one expert falls short, another may exceed expectations. And one of them will always surprise you with a perfect idea or detail you hadn't considered.

You'll quickly see which ones are strong in certain areas:

- Claude might be gentler and better at emotional tone.
- Gemini might surface facts quickly or give links.
- Grok might think more unconventionally.
- ChatGPT might give you structure and clarity.

Tip: *Let each AI play its part. They shine in different ways.*

How to Organize It

On your **phone**, open all four AI apps.
Copy your question to all four.
Rename each conversation using the *same* title—something memorable.

On your **computer**, log into the same accounts. Now your conversations are synced.

You can:

- Copy answers into a Google Doc
- Compare results side-by-side
- Highlight key points across all 4

Formatting is smoother on a computer. Lists, indents, and headings copy over perfectly. Use your phone for speed. Use your desktop for clarity.

Tip: *Keep your notes organized. It makes reviewing and sharing easier.*

This Method Empowers Your Entire Team

Churches, ministries, and Mutual Success Teams can use this AI method for:

- Filling out funding applications
- Planning community outreach events
- Researching laws and policies

- Drafting business plans
- Preparing for presentations or meetings

Instead of relying on one voice—or one person—you're building a digital council of advisors. You're not outsourcing the decision. You're *multiplying your wisdom.*

Tip: *You're not replacing your team—you're backing them up with intelligence.*

Use What Makes Sense for You

You don't have to use all four if it feels like too much.

Even using *two* gives you an advantage over just one.
Even using *one* gives you an advantage over guessing.

This is not about being perfect. This is about being *prepared.*

The Team Success Network is about multiplying strength—spiritually, relationally, and strategically. This is just one more way to walk in that principle.

Tip: *Start where you are. But don't stay small. Multiply the insight you receive.*

Final Word: Your Digital Team of Experts

You don't need to spend money. You don't need to overthink it.

You now have free access to some of the most advanced minds on Earth—trained to serve you answers 24/7. Use that. Ask big questions.

Compare their responses. Write better. Lead stronger. Understand more.

And always—let the Holy Spirit lead above all.

AI doesn't replace the Spirit. It just helps the Church catch up to the pace God's already moving at.

Tip: *Lead with wisdom. Multiply with insight. And always stay Spirit-led, even when using tech.*

Chapter 8 – The Bible Chapter Coach: Letting AI Guide Daily Word-Based Encouragement for Your Team Success Walk

Using AI to Discover the Right Bible Chapters for Growth, Clarity, and Team Encouragement—Every Single Day

---

A New Way to Study the Bible—Together and on Demand

Imagine if you had a Bible teacher who was available 24/7, could point you to the right Scriptures instantly, and could adapt your study based on your situation. That's what we now have access to through AI—**a tool that functions like a Bible Chapter Coach**, helping you explore entire passages of the Word of God, based on your needs, goals, or current season.

This is a spiritual breakthrough. Why? Because many believers don't know *where* to start in the Word. Others only get a verse or two when they really need a *chapter* that speaks deeply to their heart. And if you're leading or discipling others, trying to equip your Mutual Success Team, or planning a study—you used to need to call a pastor or wait for help. Not anymore.

Now, you can open an AI tool like ChatGPT or Claude, and type:

"Give me 5 full Bible chapters that help with trusting God through financial stress."
Or:
"What are 3 chapters that will encourage our team before launching our outreach?"

The results? Practical, relevant, and rooted in the Word. And they're free.

Why Bible Chapters (Not Just Verses) Make a Big Difference

There's nothing wrong with reading single verses. But when you're looking for depth, context, and transformation, **reading whole chapters is more powerful**. A single verse gives you a glimpse. A chapter gives you the full picture. It's like seeing the whole scene instead of one sentence of a conversation.

When AI serves as your Bible Chapter Coach, it doesn't just give you a string of disconnected Scriptures. It can guide you into full chapters that walk you through a theme, a character's journey, or God's process from problem to promise.

Examples of great prompts include:

- "List 4 Bible chapters that show God's faithfulness through long delays."
- "What are the best chapters for someone starting a faith-based business?"
- "Give me chapters for team encouragement during a tough season."

- "Help me find chapters that show unity among believers or churches."

These are spiritual, team-centered, and highly relevant. And with these prompts, you don't need to be a scholar—you just need to be curious, prayerful, and willing to ask.

How This Helps Your Mutual Success Team Walk in the Word Daily

The Bible Chapter Coach is not just a personal growth tool. It's a team-building weapon. As your Mutual Success Team grows together, you'll need shared encouragement, shared values, and shared revelations from Scripture. AI helps by offering:

- **Daily or weekly chapter lists** tailored to your team's needs
- **Bible study starters** for your next Zoom or in-person session
- **Scripture-based emails or texts** that reinforce your shared purpose
- **Topical Word plans** that walk your group through faith, abundance, unity, or vision

Here's an example: Let's say your team is preparing to launch a community garden. You can prompt:

"What are 3 Bible chapters that show the blessing of sowing and reaping, team effort, and trusting God with the harvest?"

Now your team isn't just doing something good. They're **anchoring it in the Word**, together. That creates unity, spiritual depth, and confidence—because the Bible is involved from day one.

Turning AI Recommendations into Full Bible Courses

Once you begin prompting AI for Bible chapters by topic, you'll quickly discover that you can **build entire courses** around the insights you receive. What starts as a list of 5 chapters can become a 5-day or 5-week study. Add a few discussion questions and prayer points, and you now have a complete resource for your team—or even your church.

Here's how to do it:

1. Prompt AI: "Give me 5 Bible chapters on building with excellence."
2. Follow up: "For each chapter, write a 1-paragraph summary and a discussion question."
3. Then ask: "Now turn this into a weekly course outline for our Mutual Success Team."
4. Bonus: "Suggest a prayer focus and action step for each session."

You now have a Spirit-led course that's ready to teach.

Later in this book, we'll show you how to **save that content into a free Google Doc**, format it well, and share it with others—so your Bible learning multiplies beyond your group. You can even send it to the Team Success Network and contribute to a global movement of Word-based training.

This Is the Coach You Didn't Know You Needed

We're not saying AI replaces your pastor. Or your study Bible. Or the Holy Spirit. But we are saying this: **AI can now help you quickly access the parts of the Bible that speak directly to what you or your team are facing right now.**

In a world of noise, confusion, and distraction, this is a quiet revolution. You no longer have to scroll endlessly or hope you land on the right verse. You can ask wisely, receive instantly, and spend your time **reading**, not just searching.

And as you read the Word, something supernatural happens.

- Confidence rises.
- Discernment sharpens.
- Unity deepens.
- Vision grows.

That's what happens when your team walks in the Word together.

Final Word: Equip Yourself in the Word—Every Day

The Bible says, *"Man shall not live by bread alone, but by every word that proceeds from the mouth of God."* (Matthew 4:4). That's still true today. But now you can access that "Word" faster, clearer, and more consistently—with the help of your AI Bible Chapter Coach.

Use it for your own study. Use it for your team's growth. Use it for creating studies, devotionals, and training materials. It's not cheating—it's stewarding.

Let AI help you find the chapter. Let the Holy Spirit bring the power. Let your team walk in the Word with joy and clarity, every single day.

This is how we build up the Church. This is how we grow successful teams.
One Spirit-led chapter at a time.

Chapter 9 – From Idea to Income: How Mutual Success Teams Use AI to Build Cash-Flowing Businesses

Turning God-Inspired Projects Into Financial Breakthroughs With Strategy, Tools, and Teamwork

---

Why Christian Teams Should Build Businesses Together

We've been trained to think ministry and money live in separate houses. But in the Kingdom of God, they're neighbors. God not only provides through miracles—He also provides through **ideas, businesses, and partnerships that produce ongoing income**. That's why Mutual Success Teams are not just prayer groups or outreach squads. They are meant to be **engines of transformation—including financial transformation.**

There's no shame in generating income. There's power in it. When your team has reliable cash flow, you can support families, fund outreaches, meet needs in your city, and sow into the next church. You're not stuck waiting on donations—you're actively producing Kingdom resources.

That's where AI becomes your best business partner. Not the source of blessing—but a tool that helps you move faster, think clearer, and organize smarter. AI helps you go from **inspiration to implementation. From dream to draft. From calling to cash flow.**

And when your business project is done as a team—sharing time, wisdom, and profits—the result is not just success. It's **Mutual Success.**

How AI Turns Business Dreams Into Launch-Ready Projects

A lot of believers have a dream for a business. Fewer have a plan. That's where AI steps in. What used to take months of trial and error can now take days—or even hours—when you use AI to build your business like a Spirit-led blueprint.

Here's how Mutual Success Teams are already using AI to launch real business projects:

- **Idea Generation:** "Give us 10 income-producing business ideas that fit our church's skills and values."
- **Business Naming:** "Help us brainstorm names for a Christian deck company focused on encouraging families."
- **Market Research:** "Who are our likely customers? What problems are they facing? What are they already buying?"
- **Product Planning:** "Give us a product development checklist for a physical product with printing and shipping."
- **Content Creation:** "Write 5 email templates and 3 social media posts to promote our launch."
- **Sales Strategy:** "Give us 10 low-cost ways to sell this product as a team, with or without a website."

You still bring the prayer, the direction, and the mission. But AI gives you structure, speed, and systems—so you can stop talking and start launching.

The Power of Launching a Business as a Team

In traditional business, you're expected to figure everything out on your own. But in a Mutual Success Team, you don't have to be the expert at everything—you just need to bring your part.

When you build together, something powerful happens:

- One person has the design vision
- Another handles outreach and messages
- Another handles finances and strategy
- Another builds a list or website
- The rest pray, promote, and fulfill orders

And just like that—you have a team-based, Spirit-led microbusiness with purpose and power.

Even more, this kind of business is not about getting rich. It's about **creating sustainable income while advancing the Gospel**. The more your team succeeds, the more you can give, help, and sow into others. That's why building businesses with AI and teamwork isn't a distraction from ministry—it's a direct extension of it.

Your First Project: Christian Decks (Bible Scripture Playing Cards)

Let's get specific. The Team Success Network has created a "first official project" for Mutual Success Teams who want to generate income quickly, with a product that's already been tested, designed, and spiritually impactful. That project? **Christian Decks**—a series of beautifully designed Bible Scripture Playing Cards.

Each deck contains scripture-based affirmations, faith-building content, and custom illustrations. They're tools for evangelism, encouragement, discipleship, and gifting. And they're also one of the easiest ways to start a business together as a church or group.

Here's why it works:

- The product is ready (or nearly ready)
- You don't need to invent something new—you just choose how to sell it
- AI can help you craft the emails, ads, website text, and video scripts
- Every team member can take a small role—and see big results
- The potential income goal? $10,000/month and beyond—with the right sales channels

And remember: this isn't just about one deck. Once your team understands how to sell one, you'll have the blueprint to sell many. You can customize decks for youth, marriage, parenting, healing, or outreach. It becomes a system. A stream. A movement.

What AI Can Do for Your Christian Decks Project

Here's how to prompt AI to help you build your first Christian Decks business as a Mutual Success Team:

- "Write a mission statement for a Christian product that helps families grow in the Word."
- "Give us 7 promotional messages we can use to launch our Bible card deck on social media."
- "Help us brainstorm 10 low-cost methods to sell this product through churches, pop-ups, or small groups."
- "Draft a sample product description for our store page or order form."
- "Create an outline for a one-page website that introduces the deck and invites people to buy."

With those prompts, your team will go from concept to campaign—faster than you thought possible.

And the best part? The content is tailored to your voice. You can prompt AI to speak in a friendly, faith-filled, Scripture-rich tone—so everything you release sounds like it came from *your* heart.

Final Word: Start with a Win-Win Project—Then Multiply

You don't need to start from scratch. You just need to start. Christian Decks is your open door. It's a tool for discipleship. A product for outreach. A bridge for income. And a perfect first Mutual Success Team project that can demonstrate to everyone that **faith and business were never meant to be separate.**

Once you've proven the model—just one deck, just one launch—you can take what worked and **multiply it**. Share your lessons.

Document your systems. Use AI to help write the case study. Then help another Mutual Success Team do the same thing.

That's how you move from idea to income.

That's how you create sustainable, Spirit-led cash flow.

That's how the Church becomes an unstoppable force of provision, purpose, and progress.

Part 3: The Plan – To Use AI – Leveraging Existing Blessings & Tools – To Our Christian Advantage

This part takes vision and turns it into action. The focus shifts to creating intelligent, practical plans that can be carried out by any "Mutual Success Team." Readers are guided on how to start with an idea, use AI to build it faster, and scale it into something sustainable and powerful.

By tapping into AI's research, planning, and documentation capabilities, teams gain the tools to develop solid business and ministry models that reflect God's heart and their community's needs. These are not just ideas—they are executable plans that can generate fruit quickly and consistently.

Everything here is grounded in the concept of stewardship. AI is not used for vanity—it is used to serve others. Whether through refining strategies, automating parts of communication, or structuring outreach efforts, this section reveals just how far teams can go when guided by a Spirit-filled plan.

By the end of this part, "Team Success" stops being theoretical. It becomes practical. It becomes real. Readers now have tools, steps, and a clear roadmap—backed by AI—to walk forward in faith and build the vision God has placed in their hearts.

Chapter 10 – Building Smarter Projects: How Mutual Success Teams Use AI to Plan and Launch Faster

Using AI to Accelerate Kingdom Projects with Clarity, Teamwork, and Strategic Execution

---

Speed Matters When the Mission Is Urgent

Let's be real: many Christian teams have great ideas—but they stall in the planning phase. They talk, they brainstorm, they pray... and then weeks pass. Nothing is written down. No one knows the next step. The urgency fades. And the opportunity slips.

But what if it didn't have to be that way? What if you could take a vision from your heart and have a draft plan ready in 10 minutes? What if you could launch in days instead of months—without losing spiritual depth?

That's what AI now makes possible. When used by a Spirit-led team, AI becomes a "planning partner" that helps you move faster *with focus*. It doesn't replace your values—it reflects them. It doesn't override your prayer life—it supports your obedience.

And when your Mutual Success Team uses AI the right way, you don't just build faster—you build smarter.

You still hear from God. You still use wisdom. But you also **use the tools available** to act quickly when doors open. That's how we redeem technology—and redeem the time.

Why Teams Get Stuck—and How AI Removes the Bottlenecks

Here's what slows down most Kingdom projects:

- No one wants to write the first draft
- Too many ideas, not enough clarity
- People are busy, so follow-up slips
- The plan is too vague to assign real tasks
- The tech side feels overwhelming

But when AI enters the room, these bottlenecks break wide open. For example:

- AI writes the **first draft** of your project plan
- AI summarizes brainstorming sessions into action items
- AI suggests timelines, roles, and checklists
- AI formats your plan into a Google Doc or email
- AI reminds your team what the next step is

Now, instead of waiting on one person to "get to it," you have **momentum within the hour.** And once people see the path clearly, they step up. Clarity creates movement. And movement builds success.

The Project Blueprint You Can Prompt in Minutes

Here's a sample process you can use—today—with your own Mutual Success Team. All you need is a free AI platform and a bit of direction from the Holy Spirit.

Prompt:

"Help us create a launch plan for a Mutual Success Team outreach project called 'Saturday Family Food Market,' where we provide groceries, prayer, and Gospel encouragement in a church parking lot."

AI will return something like:

- A 3-week prep timeline
- A volunteer list with role descriptions
- Suggested promotion methods
- Follow-up strategies for attendees
- Budget items and a donation plan

You can then ask:

"Turn that into a checklist in bullet format."
"Write the flyer text and announcement."
"Suggest Bible verses to encourage the team each week."

And just like that—your planning document is ready. What would've taken 4 weeks of scattered emails just took **40 minutes of guided prompting**. Your Mutual Success Team is no longer stuck. You're launching.

Start With the Plan—But Let the Spirit Breathe On It

AI isn't prophetic. It doesn't know what God is saying to your group. But it does know how to take *what you've already heard from God* and structure it into something others can follow.

This is the part where Spirit and structure meet.

When your Mutual Success Team hears from the Lord in prayer, you don't have to wonder, "What do we do now?" You can simply ask AI,

"Write a simple plan for what we just sensed in prayer."

And then **refine it** together. Adjust what's needed. Add what's missing. Insert Scriptures. Assign tasks. Pray again. Then execute.

This process doesn't remove the Spirit—it multiplies the obedience. Instead of having a few people carry the load, now everyone sees the big picture. Everyone knows their part. Everyone moves.

Smart Project Categories to Launch With AI Support

Here are examples of projects that AI can help your team plan, build, and launch:

1. Events & Gatherings

- Outreach nights
- Prayer breakfasts
- Conferences or pop-ups
- Healing rooms or community dinners

2. Business Launches

- Christian Decks sales teams
- Marketplace outreaches
- Online stores for Christian products

3. Training Programs

- Volunteer onboarding
- Team-based Bible studies
- Discipleship challenges

4. Media & Communication

- Newsletter campaigns
- Social media series
- Testimony compilation projects

You don't need to build all these at once. Start with one. Ask AI to help. Let it carry the structure—so your team can focus on faith, love, and action.

How to Keep the Team Synced as You Build

Even with a great plan, teams fall apart if they don't stay coordinated. Here's how AI helps your Mutual Success Team stay in sync:

- **Daily check-ins** using auto-generated reminders
- **Progress dashboards** that AI updates based on meetings
- **Team summaries** after every planning session
- **Volunteer messages** that are clear, warm, and Spirit-aligned
- **Shared documents** hosted in Google Drive or Notion with AI organizing the folders

You still need real people. You still need communication. But now, with AI, everything gets clearer—faster. Fewer missed steps. Less burnout. More success.

And the best part? **You're not doing more work. You're doing smarter work.** Work that flows. Work that breathes. Work that gets done.

Final Word: Don't Just Build Big—Build Smart

You don't need a megachurch budget or a Silicon Valley team to launch a God-inspired project. You just need a team that prays, plans, and **uses the tools available.**

AI is one of those tools. It won't give you your assignment—but it will help you deliver on it. It won't provide your passion—but it will reduce your friction. It won't replace the power of the Spirit—but it will help you act on what the Spirit is already saying.

So if your Mutual Success Team has been sitting on a dream—**don't wait another month.** Open a laptop. Gather the team. Inquire of the Lord. Then prompt the plan.

And before you know it—you won't just have a vision.

You'll have a launch.

Chapter 11 – "No Need Among Them": Using AI Searches to Uncover Real Solutions, Jobs, Housing, and Help for Every "Team Success" Member

How Spirit-Led Teams Use Smart Tools to Eliminate Lack and Unlock Kingdom Provision—Together

---

What Would It Look Like to Have No Needs Among You?

In Acts 4:34, the early Church was described with a phrase that still stirs hearts today: *"There was no needy person among them."* Can you imagine that? A group of believers so well connected, so willing to share, so strategically aligned, that no one lacked food, work, shelter, or dignity?

That wasn't just a miracle. That was a model.

And now, with AI as a tool in our hands, **Mutual Success Teams can begin walking in that same reality**—on purpose. This chapter is about getting super practical: helping team members find jobs, housing, food, childcare, and other forms of support using today's search tools—but led by the Spirit, rooted in faith, and done as a team.

When someone in your group is struggling, you don't have to just say, "We'll pray for you." You can also say, "Let's search together." AI searches, when prompted wisely, can uncover local, real-world help that changes lives.

We're not talking about vague theory. We're talking about practical searches that lead to solutions.

Why Searching With AI Is Better Than Searching Alone

Anyone can go to Google. But not everyone knows how to search with purpose. And most people don't know how to turn that search into a breakthrough. That's where AI tools like ChatGPT, Claude, or Gemini come in.

Instead of just looking for websites, **AI interprets your request**, summarizes the most useful info, and gives you direct actions to take. But more than that—it lets you ask follow-up questions like, "What are the risks?" or "How should I approach this conversation?" or "Can you write me an email to send to this program?"

Here's how AI helps solve real needs faster and more gently:

- Filters through the noise and gives clean summaries
- Finds *local* or *relevant* results based on what you input
- Offers strategy—not just info
- Writes scripts, emails, or applications for you
- Suggests what steps to take first

- Encourages follow-up and tracking so people don't fall through the cracks

In a Mutual Success Team, this becomes a shared responsibility. When someone is in need, the team rallies—and AI helps **search, guide, and organize the way out.**

Real Examples of What You Can Search With AI

These are not hypothetical. These are real prompts you and your team can use—today—to help someone in your group:

Housing:

"What low-income or faith-based housing programs are available in Houston, Texas for single moms?"
"Find transitional housing for a man just released from prison in Georgia."
"Help us write an inquiry letter for a room rental application for a quiet Christian family."

Jobs:

"List 10 flexible jobs that require no college degree and allow remote work."
"Find Christian job boards or hiring ministries in my city."
"Help me build a resume using my volunteer and ministry experience."

Help with Food or Bills:

"What churches near me offer weekly food distributions or grocery cards?"

"Find nonprofit programs that help with utilities in [zip code]."
"Help me write a respectful message to request help from a local pastor."

Support Systems:

"What are 3 trusted Christian addiction recovery programs that offer free counseling?"
"Find local services for grandparents raising grandchildren."
"Give us a 1-page summary of local help options we can hand out at our church."

The need is real. But so is the help. **And AI can be the bridge**—when your team leads with compassion, follow-through, and faith.

How to Search With the Spirit, Not Just With Logic

This kind of work isn't just technical—it's *relational and spiritual.* Before you prompt AI, take a moment and pray. Ask, "Holy Spirit, what is the real root of this need? What kind of help will bring lasting change?" Then search.

Let your prompting be clear and compassionate. Ask real questions, not generic ones. Include location, time frame, and specific needs.

Examples of Spirit-led prompting:

- "Help me write a gentle message to my brother who's been out of work and is losing hope."
- "What are Scripture-based steps to rebuilding after eviction?"

- "Give us 3 team action ideas to support someone who just lost their job."

These are not just tech searches. **They are Spirit-guided solutions.** You are asking for more than help. You are asking for healing, restoration, and dignity.

And the more specific you are, the more powerful the outcome.

Build a "No Need Among Us" Team Culture

This isn't about handouts. It's about a culture shift.

In the early Church, people brought what they had and shared what was needed. That's the culture of Mutual Success Teams. But now we have tools that allow us to **move with more speed, accuracy, and intentionality.**

Here's how to set that culture:

- Appoint a "Resource Finder" in your team who prompts AI to find help each week
- Keep a shared Google Doc of local support programs and services
- Encourage members to say what they need—with no shame
- Use AI to build a list of affordable Christian counselors, job training programs, or rental assistance agencies
- Print handouts from your AI search and make them available at church or team meetings

When this becomes normal, your group becomes a *resource hub*. Needs don't just get heard. **They get met.** And testimonies begin to roll in.

Final Word: Let's Make Acts 4:34 a Reality—One Prompt at a Time

You don't need a million dollars to meet needs. You need a willing heart, a wise team, and the right tools. The early Church had none of our tech, but they had unity, generosity, and Spirit-led action.

Now we have tools that can help us do it faster—if we use them with love and intention.

Let AI serve your compassion. Let it assist your generosity. Let it become a bridge between *problem* and *provision.*

And let your Mutual Success Team be known for this: **When someone joins your group, lack doesn't last long.**

This is Kingdom culture. This is spiritual strategy. This is how we build a world where there is truly—no need among us.

Chapter 12 – Testimony Amplification: Using AI to Share What God Has Done—With the World

Letting the Miracles Multiply by Capturing, Clarifying, and Communicating Testimonies at Scale

---

Your Story Was Never Meant to Stay Small

Every time God moves—He wants more than just the person He touched to be encouraged. **Testimonies were made to travel.** In Scripture, testimonies weren't private events—they were declarations. Stories of healing, breakthrough, provision, and salvation were told again and again... until they became the foundation of faith for whole generations.

But here's the reality: today, thousands of God-moments go unshared. Miracles are forgotten. Answers to prayer are reduced to a sentence. And the full power of a testimony never reaches the people who need to hear it.

Why? Because testimonies are often hard to capture, and harder to communicate. People struggle to write them well. Teams don't have time to format and publish them. The details get lost. The story stays small.

That's where AI changes everything.

AI doesn't replace the voice of the person—it **amplifies it**. It helps testimonies become clearer, more organized, more shareable, and more scalable. And when a Mutual Success Team commits to testimony amplification, you're not just celebrating God's work—you're multiplying its reach.

Why AI Makes Testimonies Easier to Share and Easier to Receive

When someone tries to write their own testimony, they often say:

- "I don't know where to start."
- "I'm not a writer."
- "It sounds better in my head than on paper."
- "It feels awkward to brag on myself."

These are normal. But they can be overcome.

AI can take a messy, unfiltered testimony and help it become:

- **Organized and readable** (so people stay engaged)
- **Formatted with emotion and clarity** (so it hits the heart)
- **Adjusted for different platforms** (social post, newsletter, flyer, podcast)
- **Translated into other languages** (to reach even more people)
- **Structured for video** (with intro, body, and conclusion)

And here's the beauty: the voice of the person is preserved. You can prompt AI with:

"Take this voice note and help me turn it into a powerful 3-paragraph written testimony."
"Write this in the same warm and faith-filled tone I used when I shared it out loud."

Suddenly, the story becomes a tool. A weapon. A spark for someone else's miracle.

The Ripple Effect of a Well-Shared Testimony

When a testimony is amplified through AI, it doesn't just sound better—it goes farther. Your Mutual Success Team can create a testimony ecosystem that reaches:

- **Other churches** looking for encouragement
- **People who are doubting** and need evidence of God's goodness
- **Partners and donors** who want to see the fruit of what they've sown
- **New team members** who wonder, "Does this really work?"
- **Local community members** who see transformation and come curious

Imagine this: A woman is healed during a prayer night. The story is captured on a phone, then run through AI to be cleaned up, formatted, and turned into:

- A social media post

- A blog story
- A printed flyer for future outreaches
- A video script for a 60-second reel
- A line in a newsletter to supporters
- A quote on your homepage
- A faith-building message for your next team meeting

That's one testimony, amplified seven ways.

Now it doesn't just encourage a room—it transforms a region.

How to Prompt AI to Help You Capture and Amplify Testimonies

Let's get practical. Here's how you can use AI to build a testimony system within your Mutual Success Team:

Start with these prompts:

"Take this paragraph and make it into a 3-part testimony: What happened before, what changed, and how life is now."
"Write this in a warm, emotional tone—like it's being shared at a church."
"Turn this story into a 5-slide PowerPoint testimony with Bible verses included."
"Translate this into Spanish and make it suitable for a youth outreach newsletter."
"Create a testimony summary for a printed flyer to hand out next Saturday."
"Help me turn 3 quick stories into a single 'highlight reel' for our social media."

You're not lying. You're not exaggerating. You're just making the truth **easy to tell, beautiful to share, and powerful to repeat.**

That's not marketing. That's **ministry**.

Team Roles and Systems That Help You Capture Every Miracle

Don't wait until the testimony is perfect. Set up a system that helps your team capture God's goodness in real time:

1. Appoint a "Testimony Coordinator"
This person isn't in charge of writing everything—but they make sure testimonies get collected and submitted.

2. Use Voice Notes
Encourage people to record their stories in 1–2 minutes. Then let AI summarize and clean them up.

3. Create a Submission Form
Use Google Forms to ask 3 simple questions: What happened? What changed? What's life like now?

4. Use AI to Sort and Style
AI can help categorize testimonies (healing, provision, deliverance, restoration) and rewrite them for different uses.

5. Build a Testimony Archive
Store your stories in a shared folder (Google Docs or Notion). Over time, this becomes a *library of faith*.

This system is simple. But it turns scattered memories into spiritual blueprints. **Every time God moves—you document, share, and seed another miracle.**

Final Word: Let the Miracle Echo

God's goodness was never meant to be whispered in private. It was meant to be shouted from the rooftops, printed on banners, written in newsletters, and posted in the streets. That's what testimony amplification is about.

AI doesn't replace the miracle—it magnifies it. AI doesn't add hype—it brings *clarity and reach*. AI isn't showing off. It's **pointing up**.

Your Mutual Success Team is about more than success. It's about stewardship. And that includes stewarding the stories of what God has done.

So let the miracle echo.

Write it. Format it. Share it. Translate it. Post it. Print it.

And let someone else's breakthrough become the spark for their own.

Chapter 13 – Supernatural Meets Superintelligence: Using AI to Multiply Team Success Impact

There's something new happening. It's not just technology. And it's not just revival. It's both.

We're entering a time when the supernatural power of God is beginning to work alongside the superintelligent tools of our generation. Not to compete — but to multiply each other's impact. That's the world Team Success is stepping into.

And it's where churches are going to thrive, if they're willing to let both power and precision flow side by side.

What Happens When You Combine Miracles with Momentum?

God still heals. The Spirit still moves. That hasn't changed. What's changed is what we can do **right after** the miracle.

In the past, someone would get healed and the story would stop there. Now, with a phone and AI tool in hand, that testimony can be:

- Written, formatted, and translated into 20 languages
- Sent to every partner church in your Mutual Success Team

- Posted with Scripture graphics and hashtags in minutes
- Turned into a short video with captions by AI

The same power that healed the person is now extended into **hundreds of other lives** who get to hear about it. You've just used superintelligence to multiply the supernatural.

Tip: Don't let miracles be isolated. Let them multiply.

AI is a Tool, Not a Threat

There's no war between Spirit and software. They are not on opposite sides. God gave wisdom to people. People built tools. And now those tools can help the Church advance faster — as long as we keep the Holy Spirit in charge.

The key is to never let technology replace relationship. AI can help organize your strategy, but only God can give you the assignment.

So we don't worship the tools. We use them to **serve the mission**.

Tip: Let God lead. Let AI assist.

Multiply, Document, Scale, Repeat

Mutual Success Teams thrive on duplication. But duplication takes documentation. And that's where AI shines.

Let's say a team of two churches builds a healing outreach program that results in 34 people healed in one month. That's amazing. But the next question is: **how do we share this with 10 other cities?**

Here's what you can do:

- Ask AI to document everything — schedules, roles, outreach text, FAQ, follow-up strategy.
- Translate that into a clean PDF guide with links and photos.
- Send it to every connected church in your Team Success Network.

Now what God did in one city becomes the seed of revival in ten others.

Tip: When something works, capture it. Share it. Multiply it.

AI Doesn't Heal — But It Can Help You Get There

AI can't lay hands. It can't command pain to leave. It doesn't pray in tongues.

But it can:

- Tell you where the healing rooms are
- Organize transportation for 40 people who need healing
- Write a press release to invite others
- Suggest ways to present miracle testimonies legally and clearly
- Keep track of follow-ups to make sure healed people get discipled

And that's exactly the point. When AI is used properly, it clears the path for the power of God to flow more freely.

Tip: Let the tools help set the table. Then let God show up with the meal.

Final Word: Don't Be Afraid to Use Both

There's a wave coming. Churches who use only AI will be fast, but empty. Churches who use only spiritual power may be deep, but often overwhelmed.

But those who use both — Spirit and strategy, supernatural and superintelligence — will go the farthest, the fastest, with the fewest wasted steps.

That's what Team Success is about. That's where the future is going. And it's already happening.

You don't need to choose between miracles and systems. God is asking you to use both.

Tip: When the supernatural joins with superintelligence, entire cities' needs are ministered to even more completely and comprehensively. Leveraging the full power of each gives us tremendous advantages for accomplishing our mission of bringing The Kingdom of Heaven closer.

Part 4: Life Transformation – Personal & Business – You Need Both – For Our Christian Advantage

This section explores how AI can support believers not just in ministry work, but in their everyday lives. It brings to light the personal growth and practical changes that come from having consistent, Spirit-guided access to tools that answer questions, offer encouragement, and organize daily priorities. AI becomes a daily companion for those seeking steady, step-by-step progress.

Whether the goal is managing your energy, organizing your business plan, or just staying focused on what matters, AI can offer support at any level of intensity. It's not a substitute for faith—it's a magnifier of faith in action. When the tools of this age are placed into the hands of a believer who inquires of the Lord, remarkable things begin to happen.

The blend of business transformation with deep personal encouragement is central to this part. For those who need more structure, AI offers it. For those needing healing or fresh perspective, it responds. For those ready to grow beyond where they've been, it walks beside them. In short, it helps you transform the way you live your calling.

By the end of this section, the reader begins to see that the power of AI is in its availability. You don't have to wait on a meeting or mentor—you can ask now, build now, grow now. That's how life change happens—with tools that support your faith-fueled momentum.

Chapter 14 – Building a 'Mutual Success Team' "AI Lab" – To Explore, Learn, & Use the Best Solutions You Discover

How to Create a Simple, Shared Space Where Your Team Can Explore, Learn, and Apply AI Tools Together

---

Why Your Team Needs Its Own ""AI Lab""

If you want your Mutual Success Team to operate faster, smarter, and with less burnout—you don't need more meetings. You need more clarity. You need a process that lets everyone learn, test, and apply AI tools in a safe, shared, Spirit-led way.

That's what an **"AI Lab"** is for. It's not a physical room with blinking lights. It's simply **a shared space—digital or in-person—where your team plays with AI tools, compares results, and builds solutions together.** It's where creativity meets Kingdom work. And it's where curiosity becomes strategy.

Think of it like a sandbox: a place where you can test ideas without pressure. You can try prompts, compare tools like

ChatGPT and Gemini, and document what works best. Instead of everyone learning AI in isolation, you learn as a team.

And the result? Every person on your team becomes more confident, more capable, and more valuable—because they've seen what AI can do *for them* and *for others.*

What a "Team "AI Lab"" Actually Looks Like

You don't need special software or advanced tech. Your "AI Lab" can start with:

- A shared Google Doc to record prompts and responses
- A group chat (like WhatsApp, Signal, or Slack) to share ideas
- A weekly Zoom or in-person session to explore tools
- A volunteer "AI Champion" who helps lead experiments and gather insights
- Free accounts on AI tools like ChatGPT, Claude, Grok, and Google Gemini

That's it. Nothing fancy. Just a committed team saying, "Let's try this together and see what's possible."

Each person takes turns trying a task. One writes a flyer using AI. Another drafts a training plan. Another prompts a script or vision doc. Then the team **compares the outputs, tweaks the prompts, and discusses what worked.**

This builds shared knowledge fast. And soon, everyone becomes a co-builder of systems, content, and strategies. That's when your

Mutual Success Team stops depending on one tech-savvy person—and starts **scaling success through shared learning.**

How to Start—Step-by-Step Sample Session

Want to launch your first "AI Lab" session? Here's a sample run-through you can use this week:

1. Choose a Simple Task
Let's say the goal is to "Create a one-page flyer to invite people to your Mutual Success Team launch event."

2. Break into Teams or Individuals
Each person or small group uses a different AI tool:

- One uses ChatGPT
- Another uses Claude
- Another uses Gemini
- Another uses Microsoft Copilot or another platform

3. Prompt the Same Request in Each Tool
Example prompt:

"Write a friendly, faith-based flyer to invite people to a Mutual Success Team info session. Mention food, vision, and the goal of eliminating need in our city."

4. Review and Compare Results
Bring everyone together and read each AI's output. Discuss:

- What did each one do well?
- Which sounded the most aligned with your values?
- What needs to be adjusted or improved?

5. Improve the Prompt Together
Try a second round:

"Make it shorter. Add a scripture. Keep the warm tone but more excitement."
Then compare again.

6. Finalize and Save
Pick the best version. Save it in your team folder. Now you've got a flyer template—and everyone learned something in the process.

That's one "AI Lab" session. Fast. Fun. Productive. And repeatable.

Benefits of Running Your Own "AI Lab"

When your Mutual Success Team uses an "AI Lab" structure, you unlock new levels of teamwork, confidence, and Kingdom impact. Here's what that looks like:

- **Shared vocabulary**: Everyone starts understanding prompts, formatting, and tools
- **Reduced fear**: People stop feeling intimidated and start exploring
- **Faster execution**: Projects that used to take days now take hours
- **Skill building**: Each person becomes stronger in planning, writing, organizing
- **Better decision-making**: You can test ideas before committing to one path
- **Documentation & duplication**: You capture processes that can be reused and scaled

Most importantly: **you become a team that learns together.** And that creates culture.

When your culture is "let's test it," you don't wait around. You try. You build. You adjust. And you get results.

What You Can Use Your "AI Lab" To Create

There's no limit to what you can develop inside your lab. Once you're comfortable, begin using it to build:

- **Flyers and event invites**
- **Devotional outlines or prayer guides**
- **Business plans and product launches**
- **Job descriptions and role breakdowns**
- **Training courses and challenge outlines**
- **Content calendars for outreach or social media**
- **Volunteer management plans**
- **Partnership proposals and funding drafts**

You don't need to create it all at once. Just start with what's in front of you. Every Mutual Success Team project—big or small—can be made stronger and faster when it goes through the ""AI Lab"" process.

Final Word: Build the Lab—Then Let the Lab Build You

AI won't replace your team. But it will **multiply what your team can do.**

By building a Mutual Success Team "AI Lab", you create a space where God-given ideas become real-world strategies—quickly. You reduce friction. You shorten timelines. You create documents, plans, tools, and systems that serve real people, faster than ever before.

Don't build alone. Don't guess. Don't delay.

Start your lab. Compare tools. Test prompts. Implement what works.

And then do it again.

The lab you build today could be the launchpad for your team's greatest breakthrough tomorrow.

Chapter 15 – Harnessing a Variety of AI Tools for "Team Success" Innovation

Using the Unique Strengths of Different AIs to Create Breakthroughs in Every Part of Your Mutual Success Team

Innovation Begins With Exposure to New Possibilities

If your team is only using one AI tool, you're only getting part of the picture. Just like a body needs more than one organ to function—your "Mutual Success Team" needs a variety of tools to unlock the full power of innovation.

Innovation isn't always a lightning bolt—it often begins by exploring new conversations, new insights, and new perspectives. When you start comparing the strengths of different AI tools (like ChatGPT, Claude, Grok, and Google Gemini), you start seeing what's possible—and from that, new ideas begin to emerge.

You don't need to be a coder or a tech genius. You just need to be a Spirit-led explorer. You need to say:
"What else could we do if we had better ideas, better strategies, and faster systems?"

That's what this chapter is about—opening your team to cross-tool innovation, so you're not stuck doing things the old way when a better way is right in front of you.

The Strengths of Different AI Tools—And Why They Matter

Each AI platform has its own personality, tone, and strengths. When you learn to use them together, you get a richer, more well-rounded set of options for every challenge.

Here's a general breakdown:

- **ChatGPT (by OpenAI)** – Best for structure, organization, and clarity. Great for outlines, email drafts, checklists, and summaries.
- **Claude (by Anthropic)** – Has an emotionally intelligent tone. Feels more "gentle," great for relational writing, coaching messages, and sensitive content.
- **Grok (by xAI/Twitter)** – More creative, quirky, and outside-the-box. Often suggests ideas you wouldn't expect.
- **Google Gemini** – Fast access to facts, summaries from web pages, and Google-based results. Great for searching or data-heavy tasks.

Why does this matter? Because different situations need different strengths. A "Mutual Success Team" building a business may use ChatGPT for a plan, Claude for outreach letters, Gemini for research, and Grok for creative marketing ideas.

Innovation comes when you combine perspectives. And now, you can do that—in seconds.

---

5 Areas Where a Variety of AIs Create Immediate Innovation

Let's get specific. These are five areas where using multiple AI tools can instantly bless and improve your team:

Strategic Planning
Using a variety of AIs lets you compare multiple ways to structure your event, business, or project. One tool gives a formal outline, another offers a creative twist, and a third introduces an idea you hadn't considered.

Communication & Messaging
Different AIs have different tones. Use Claude to write soft and encouraging messages. Use ChatGPT to create clear updates. Use Grok to inject humor or surprise. Together, your communication becomes relatable, effective, and human.

Problem Solving & Team Coaching
Facing internal conflict? Use Claude to write a gentle script. Use Gemini to research Christian leadership principles. Use ChatGPT to help plan a team-building activity.

Outreach & Marketing
Want to promote your "Mutual Success Project"? Use Grok for headline ideas. Use ChatGPT to write a sales page. Use Gemini to find keywords. Use Claude to write stories and testimonies.

Innovation Brainstorms
Want to level up your team or product? Prompt each tool with:
"How could we improve this idea using principles of simplicity, teamwork, and faith?"
The answers will amaze you.

How to Run an AI Innovation Session With Your Team

Want to spark new growth fast? Host an AI Innovation Hour with your "Mutual Success Team." Here's how to do it:

Step 1: Choose a Focus
Pick one area to improve: your welcome experience, project growth, fundraising, or community impact.

Step 2: Assign AI Tools
Each person uses a different AI tool to prompt the same question. For example:
"How can we improve the first 10 minutes of our team gatherings?"

Step 3: Collect the Responses
Have everyone copy their answers into a shared doc or present what stood out.

Step 4: Compare and Highlight
Ask:
- What surprised us?
- What can we try this week?
- Which idea would make the biggest impact?

Step 5: Act Fast
Pick one idea and implement it in the next 7 days. Document the results. Repeat monthly.

You'll quickly see that innovation isn't about being smart—it's about being exposed to more options, together.

How Prompting Different AIs Unlocks Different Wisdom

Try this experiment: take a prompt like
"What's a good way to help team members stay encouraged during a long-term project?"

Here's what you might get:

- **ChatGPT**: A weekly checklist with motivational quotes and reflection time
- **Claude**: A gentle letter reminding people of their value and progress
- **Grok**: A wild idea like starting "Victory Voice Memos" where each person records a 15-second praise report
- **Gemini**: A link to a study showing how encouragement boosts retention

Now you're not just guessing. You're building from multiple angles of wisdom.

You still filter everything with the Holy Spirit. But now, your foundation is wider—and your confidence to build is stronger.

Final Word: Innovation Is Obedience in Action

Sometimes we wait for a vision when God is already saying, *"Use what I've given you."*

The variety of AI tools isn't a distraction—it's a gift for Kingdom builders. Each platform carries a different strength. Together, they help you invent, improve, and multiply the impact of your "Mutual Success Team."

Innovation isn't about being futuristic—it's about being faithful to grow what God has already planted.

So build your own AI toolkit. Use the strengths of each tool. Invite your team to explore with joy. Try new prompts. Compare results. And ask the Holy Spirit to highlight what's worth acting on.

Because when you harness a variety of tools, you unlock fresh breakthroughs.
And when you build with wisdom, your team's success becomes inevitable.

Chapter 16 – "Mutual Success Team" Brainstorms—Powered by All Four AIs: ChatGPT, Grok, Claude AI, & Google Gemini

How to Create Your Own Mini Think Tank to Solve Problems, Generate Breakthroughs, and Move Forward Faster Than Ever

---

What If You Could Eliminate 90% of Your Problems This Week?

We've all had those moments: a project stalls, the team feels stuck, confusion sets in, and nothing seems to move forward. Sometimes it's strategy. Sometimes it's communication. Sometimes it's just mental fog. The result? Frustration. Delay. And energy gets drained.

But what if you could flip that?

What if, instead of being stuck for days or weeks, you could sit down for 30 minutes, open four tabs on your computer or phone, ask a single question—and walk away with **four different expert-level responses** that give you options, clarity, and momentum?

That's what happens when your **Mutual Success Team uses a 4-AI Brainstorm Session**—leveraging the strengths of ChatGPT, Claude, Grok, and Google Gemini.

It's not just helpful. It's *radically empowering.* You no longer have to guess your way forward. You simply gather the best ideas, pray over them, and act in unity.

This chapter shows you how.

Why Four AIs Are Better Than One

Every AI has a different "personality" and strength. When you learn to use all four together, it's like sitting around a table with four expert advisors—each with a different voice.

- **ChatGPT** gives you clarity, structure, and high-quality drafts
- **Claude AI** offers warmth, emotional insight, and relationship-sensitive responses
- **Grok** is wild, creative, unconventional, and sometimes hilariously surprising
- **Google Gemini** brings research-backed facts, web-based summaries, and straight-up data

When you're stuck on a problem, one AI might give you a basic solution, another gives you the *why*, and a third gives you a better idea altogether. The fourth might show you what you completely missed.

This isn't overkill. This is strategy.

It's a Kingdom-minded approach to brainstorming: **You don't move forward with one guess—you choose from wisdom.**

How to Set Up Your Four-AI Brainstorm Lab

You don't need fancy software or paid accounts to make this work. Here's how to set yourself (and your team) up for success:

Step 1: Create a Free Account on All Four Tools

- ChatGPT (chat.openai.com)
- Claude AI (claude.ai)
- Google Gemini (gemini.google.com)
- Grok (accessed through X.com with a premium account, but alternatives are emerging)

Step 2: Use the Same Prompt in All Four Tools
This is key. Ask the exact same question in each chat window. Example:

"What's a smart, low-cost way for our Mutual Success Team to generate $2,000/month using our combined skills?"

Step 3: Compare the Results Side-by-Side
Copy all the responses into a shared doc or notepad. Highlight what each AI got right, what stood out, and what needs tweaking.

Step 4: Combine the Best Ideas
You'll end up with a master plan that includes:

- Creative twists from Grok
- Step-by-step logic from ChatGPT
- Sensitive phrasing and encouragement from Claude

- Extra research and links from Gemini

You've now created a **think tank without hiring a consultant.**

What Types of Problems This Can Solve

This method works for almost anything your Mutual Success Team faces. Here are a few categories where it can eliminate stress and create immediate movement:

1. Strategy Blocks
Can't figure out how to launch your project? Ask all four AIs for a launch timeline. Combine their answers. You'll be moving in an hour.

2. Conflict Resolution
Need to resolve a team tension? Ask Claude and ChatGPT to draft a kind, peace-centered message. Compare tones. Use the best.

3. Financial Planning
Need quick income? Prompt:

"Give me 5 business ideas that a church group can launch using less than $500."

Now compare feasibility across AIs.

4. Marketing or Outreach
Need a flyer, video script, or email? Ask all four AIs:

"Write a compelling message for our Christian Decks launch that inspires trust and action."

Use what fits best.

5. Spiritual Growth Resources
Want a Bible study plan? Ask each tool:

"Give us a 4-week study on faith and financial stewardship for a Mutual Success Team."

Each one will give you a unique approach. The Spirit will confirm what to use.

How to Make This a Weekly Team Habit

Don't keep this power to yourself. Turn it into a **team-wide practice**.

Create a Shared Prompt Document

- Start with a question each week: "How can we improve ___?"
- Each team member chooses one AI tool
- Everyone runs the same prompt
- Share results in a team Zoom or meeting
- Vote on best idea
- Implement within 7 days

Store Wins in an AI Archive Folder
Over time, your team will build a treasure chest of:

- Flyer templates
- Outreach messages
- Training outlines
- Prayer guides

- Business strategies
- Conflict resolutions

All searchable. All shareable. All Spirit-aligned.

What Makes This Spirit-Led and Not Just Smart?

The AI tools are just that—**tools.** What makes this process powerful is that you **start with prayer, listen for direction, and filter results through discernment.**

Before you prompt, ask:

"Lord, what question do You want us to ask this week?"
Then use the AI tools to help answer that question clearly and practically.

After you gather results, ask:

"Holy Spirit, which direction aligns with Your heart and timing?"

This way, you don't follow the AI—you **follow God's voice and use the tools to take obedient action.**

You're not replacing faith—you're accelerating follow-through.

Final Word: This Is Your New Power Move

When you're overwhelmed, stuck, or just out of ideas—don't sit in silence. **Open your "AI Lab".** Ask all four. Compare. Discern. Act.

You will be shocked how many problems melt under the light of clear thinking, fresh strategy, and Spirit-led prompting.

This isn't about tech. It's about traction.
This isn't about automation. It's about *activation.*
This isn't about being smarter. It's about being *more faithful* with the tools God has placed in your hand.

Use the four AIs. Build your brainstorm sessions. Solve your problems faster. Launch your projects sooner. And watch your Mutual Success Team start running at a speed that **blesses everyone involved.**

The breakthroughs are already there. You just need to prompt them.

Chapter 17 – Team Success Decision-Making: How AI Can Sharpen Any Strategy

Using AI for Daily Refinement, Smarter Execution, and Long-Term Growth in Your Mutual Success Team

---

Your Team Doesn't Just Need a Strategy—It Needs a Better One

Let's be honest. A lot of teams don't fail because they lack vision—they fail because their *strategy* isn't working. Or more accurately... it *was* working, five years ago. But now? It's outdated. Clunky. Bloated. Overcomplicated. Or just too slow.

That's where most Mutual Success Teams get stuck. The desire is there. The mission is clear. The people are willing. But the actual methods? They need help.

That's where AI steps in—not to replace your plan, but to help **refine your plan.** Whether you're working on outreach, business development, content creation, or volunteer coordination, AI can help you take what you're already doing... and make it faster, smarter, and more focused.

This chapter is all about **sharpening your strategies**—not by starting from scratch, but by using AI to continually improve what's already in motion. Think 1% better today, 1% better tomorrow. That's not hype. That's momentum.

The Power of CANI: Constant And Never-Ending Improvement

There's a principle from high-performance leadership known as **CANI**—Constant And Never-ending Improvement. It's famously used by Japanese automakers like Toyota, where even a 1% improvement each day results in **over 360% total improvement per year.**

Now imagine applying that to your Mutual Success Team's strategy.

- What if your follow-up system was 1% better each week?
- What if your outreach events were 1% easier to prepare every time?
- What if your income streams, planning processes, and communication flow all got 1% tighter, clearer, smoother—every single month?

That's what AI enables. It helps you move out of maintenance mode and into *momentum mode*. Because when you commit to improvement—AI gives you the **ideas, structures, and suggestions** that keep it going.

How to Use AI to Improve an Existing Strategy

You don't need to reinvent the wheel. Just bring your current strategy to the table, and ask AI:

"How can we improve this?"

Here's an example. Let's say your team runs a monthly outreach. You might tell AI:

"Here's our current outreach plan: 1) Schedule the venue, 2) Send flyers, 3) Set up chairs, 4) Pray and minister, 5) Follow-up emails. What would you improve?"

AI might suggest:

- Add a reminder system to double attendance
- Use a QR code for easier follow-up
- Streamline volunteer roles
- Add a 60-second testimony at the beginning
- Insert a donation prompt or next-step offer

You then ask:

"Rewrite this as a sharper 5-step strategy with a timeline and checklist."
Or:
"Help me train a volunteer to run this plan without me."

Now your event just became *scalable*. You've taken a good idea—and made it excellent.

Where Strategy Bloat Slows You Down—and What to Fix

Most teams suffer from "strategy bloat." That means there's too much happening, too many steps, or too many decisions being made every week that could be automated, clarified, or eliminated.

Here are common signs of bloat in your Mutual Success Team:

- You're writing the same emails every month from scratch
- Meetings run long because no one knows the action steps
- Follow-up slips through the cracks because the system is unclear
- Good ideas are never implemented because no one owns them
- Tasks get duplicated or forgotten

With AI, you can do a full audit. Just ask:

"We're overwhelmed. Can you look at our 7-step process and suggest a way to simplify it?"

Or even:

"We're running this 4-week challenge. Help us make it more efficient, with clearer next steps."

AI is like having a consultant who doesn't charge you—**but always gives you fresh insight**.

Examples of AI-Driven Strategy Sharpening for Teams

These are real examples of how Mutual Success Teams are sharpening their systems through AI:

- **Email Follow-Up Strategy:** Prompt: "Help us write 3 follow-up emails after an outreach night—friendly, clear, and full of encouragement."
- **Volunteer Management:** Prompt: "Create a role description, weekly task list, and a 10-minute training script for our new project lead."
- **Outreach Systems:** Prompt: "Turn this monthly outreach plan into a repeatable model with reminders, social posts, and a preparation checklist."
- **Financial Clarity:** Prompt: "Suggest 5 ways we can generate $1,000/month as a team, using our existing assets and skills."
- **Time Stewardship:** Prompt: "Help us reduce our weekly meeting from 90 minutes to 45—without losing momentum."

You're not guessing. You're testing. And each time you sharpen a piece of your system, your entire team gets stronger.

Why the Holy Spirit Still Leads—And AI Helps Us Follow More Effectively

Let's make this crystal clear: AI doesn't replace revelation. God gives you the vision. The Spirit speaks to your heart. But once He does—**you still need a system**.

Moses had a vision. But he needed a structure to lead the people. Paul had encounters with Jesus—but still asked for support, planned routes, and wrote letters with precision.

AI helps you move from vague intentions to **clear assignments and better execution.** It helps you act on what God said—not six months later, but today.

That's why refining your strategies isn't carnal. It's Kingdom.

God gets glory when your plans are excellent, your follow-through is strong, and your team feels empowered—not exhausted.

Final Word: Make Your Strategy Better, Every Week

You don't need to fix everything at once. Just commit to one thing:

Make one part of your system better this week.

That might mean:

- Rewriting your team's onboarding email
- Simplifying your outreach checklist
- Delegating with clarity using an AI-generated script
- Reviewing your business flow and asking, "What's not working anymore?"

Then prompt AI, ask for help, test the improvement—and share what worked.

When every team member adopts this mindset—**"We improve together"—your Mutual Success Team becomes unstoppable.**

It's not about overhauling. It's about optimizing. And when your team walks in daily improvement, led by the Spirit and assisted by wisdom… your results become **multiplied, sustainable, and joyful.**

Start today. Let AI help you sharpen what God has already called you to build.

Chapter 18 – Teaching Your Team to Use AI Tools Together, in Unity

How a Spirit-Led, AI-Empowered Team Moves Faster, Builds Smarter, and Breaks Through More Often

---

Why Your Team Needs to Use AI Together—Not Just Individually

One person using AI is helpful. But a whole **Mutual Success Team** using AI together? That's *powerful.* That's acceleration. That's the difference between a car going 30 mph… and a train moving full-speed with every car locked in sync.

Here's the truth: the more people on your team who know how to use AI effectively, the faster your ideas turn into action. Obstacles get overcome in hours instead of weeks. Messages get written while others sleep. Plans get outlined in minutes. And when everyone is involved—**no one is left behind.**

But this only works when you **teach your team to use AI tools together.** Not just casually, but intentionally. Not just as an individual skill, but as a **shared advantage.** That's the purpose of this chapter—to show you how to get your whole team speaking the same "AI language," building together in unity, and setting new records for Kingdom impact.

Because a united team, using AI in sync, becomes *unstoppable.*

The Difference Between One AI User and a Unified AI Team

Let's imagine two scenarios.

Scenario A: One person on your team knows how to use AI. They do the heavy lifting—writing the emails, creating the checklists, building the plans. Everyone else waits on them. That's better than nothing—but it's slow, and it creates a bottleneck.

Scenario B: Every person on your team knows how to prompt. One creates an outreach plan. Another writes the announcements. Another formats the testimonials. Another builds the social posts. Another improves the training guide.

The whole team becomes **productive, powerful, and proactive.**

Instead of waiting, people contribute. Instead of bottlenecks, you get breakthroughs. Instead of depending on one leader, the whole team rises up. That's what happens when you stop working *alone* and start working *in unity*—with AI as your shared support tool.

How to Train Your Team to Use AI—Even if They're Nervous or New

Many people hesitate to use AI tools because they feel overwhelmed or assume they're "not techie." But here's the truth: if you can text a friend, you can use AI.

Training your team can be simple and fun. Here's a suggested 3-step rollout you can use this week:

Step 1: Set Up Accounts Together

- Have a 30-minute "AI Setup Session"
- Help each person create free accounts for:
    - ChatGPT (chat.openai.com)
    - Claude (claude.ai)
    - Google Gemini (gemini.google.com)
- Bookmark the sites on their phone or laptop
- Open all four on your device to model how it works

Step 2: Run a Practice Prompt
Have everyone copy and paste the same simple prompt:

"Give me 3 outreach ideas a Christian Mutual Success Team could do in the next 30 days with no budget."

Let everyone compare the results. Smile. Laugh. Celebrate the power.

Step 3: Assign a Team Challenge
Give everyone a task for the week using AI:

- Write a thank-you note
- Draft a flyer
- Make a checklist
- Prompt a devotional idea
- Summarize your project in 3 bullets

Then gather next week to share. Boom—your team is now trained, confident, and ready to build together.

Why This Builds Unity and Momentum at the Same Time

When every team member is equipped to use AI, something shifts. You no longer have spectators—you have **builders.** You don't have people waiting to be told what to do—you have people offering solutions.

And that unity? It's not just practical—it's *spiritual.*

Psalm 133 says, *"How good and pleasant it is when God's people live together in unity... for there the Lord commands a blessing."* When your team is aligned—not just in heart, but in ability—that's where the commanded blessing shows up.

And practically speaking, this builds:

- **Shared language** (everyone understands prompting, refining, reviewing)
- **Shared speed** (less waiting on others, more real-time creation)
- **Shared creativity** (new ideas fly around the room like sparks)
- **Shared ownership** (people feel included, not sidelined)

This is how you shift from **a great idea** to a **movement of momentum.**

Encourage Your Team to Build—Then Share the Results

Once your team is trained and practicing AI together, you'll start seeing results quickly. And we want to hear about it!

Here's how to share your story with the broader network:

- When you launch a flyer written with AI—**screenshot it.**
- When your team outlines a project in 10 minutes—**record it.**
- When someone creates their first testimony or training doc—**celebrate it!**
- Then email it to: **results@teamsuccessnetwork.com**

Tell us how AI helped you go faster, build smarter, or break through something that used to feel stuck. Your testimony will encourage others—and it might get featured in future resources.

Your win can spark someone else's breakthrough.

Final Word: Unity + Tools = Team Success

Don't settle for slow progress. Don't build with one builder and four watchers. Don't delay breakthroughs because people "don't know where to start."

Train your team. Equip every voice. Let every person prompt.

This isn't about replacing people. It's about empowering them.

When everyone in your Mutual Success Team uses AI together, you don't just get more done—you get it done with unity, with joy, and with confidence.

So take the leap. Host the training. Run the challenge. Share the wins.

And let your team be known not just for success—but for building in unity, with power, at record speed.

Chapter 19 – The AI "Life Coach": Supporting Your Daily Decisions and Progress in Your Team Success Journey

How to Get Encouragement, Direction, and Real Results by Letting AI Support Your Next Step—Without Ever Taking the Lead

---

You Don't Need to Do This Alone

Let's face it—doing what God called you to do can feel overwhelming. You've got goals. You've got assignments. You've got people depending on you. But between all that and your own life's needs... it's easy to feel stretched thin.

What if you had a coach on call, 24/7? Someone who could help you plan your day, sort through tough decisions, organize your thoughts, or give encouragement when you feel like quitting?

You do now. That's what AI can be—not your Lord, not your leader, but **your Life Coach.**

When you prompt AI with purpose, and with the right guardrails, you get real value:

- Direction when you're unsure
- Encouragement when you're tired
- Clarity when you're scattered
- Momentum when you're stuck

And that helps you move forward—not just personally, but in your **Team Success journey.** Because when you're strong, focused, and refreshed—you show up better for your team.

What the AI Life Coach Can (and Can't) Do

Let's get one thing straight: **AI is not the Holy Spirit.** It doesn't know your future. It doesn't carry peace. It doesn't reveal your identity in Christ.

But it *can* help with:

- Organizing your thoughts
- Reframing a tough situation
- Offering structure for a goal
- Giving 5 options you hadn't considered
- Creating a plan when you feel frozen
- Offering consistent, kind, and structured advice

AI is like a mirror—it reflects what you feed it. If you prompt it with faith, goals, and clarity, it gives back encouragement, strategy, and structure. But if you ask it vague questions—or hope it replaces your prayer life—you'll get distracted or even discouraged.

That's why this chapter is all about **training your AI to serve you well**—and never letting it run the show.

5 Tips for Getting the Most From AI as Your "Life Coach"

Tip #1: Define Your Values First
Before you ever prompt, tell the AI what matters most to you.
Example:

"Please respond in a tone that's faith-filled, encouraging, and Biblically sound. I'm part of a Christian team building ministry and business projects. Keep your advice aligned with faith, family, integrity, and clarity."

This sets the tone. You'll be amazed how differently the AI responds when you *set the frame* upfront.

Tip #2: Be Specific About What You Need
Don't just say, "Help me." Instead say:

"Help me break this goal into 5 steps I can do in the next 7 days."
"Help me evaluate 3 options I'm considering."
"What are 3 ways I can process discouragement today with Scripture and action?"

Clarity in = clarity out.

Tip #3: Prompt for Encouragement, Not Just Strategy
Sometimes you don't need a list—you need hope.
Prompt:

"Write me a short message as if you were a Christian life coach encouraging me to keep going after a tough week."

These moments remind you that *you're not crazy—you're just growing.*

Tip #4: Use It to Track and Reflect
Each morning, prompt AI:

"Here are my 3 priorities today. Help me organize them."
Each evening:
"Help me reflect: What went well? What felt off? How can I reset for tomorrow?"

It's like a journal that listens, summarizes, and supports you daily.

Tip #5: Let It Help You Support Others, Too
You're not the only one who needs a boost. Prompt AI to help you write support messages for teammates:

"Write a short encouragement message for someone struggling to meet their goals."
"Give me a prayer to send to a teammate who feels stuck."

That's how you extend your Life Coach advantage—to your entire Mutual Success Team.

The Life Coach That Never Sleeps—And Always Has Time for You

One of the most empowering things about using AI this way is that **you never have to wait**. You don't need to book a session. You don't need to wait for a callback. You don't have to explain your backstory five times.

You open your laptop. Or your phone. You type one clear, intentional question. And the coach is ready.

And here's what happens over time:

- You make faster decisions
- You second-guess yourself less
- You build consistent habits
- You track your wins
- You show up for your team better

Why? Because you're getting daily traction—not waiting for breakthroughs, but building them one step at a time.

And in a Kingdom context, that matters deeply. When we steward our time and energy well, we carry more grace, joy, and peace into every team project, every outreach, every meeting.

Sample Prompts to Use With Your AI Life Coach

Here's a few you can try right now:

For Motivation:

"Remind me why my calling matters when I feel tired."
"What's a 3-sentence pep talk from a Christian life coach perspective?"

For Planning:

"Organize my 5 priorities today in a 2-hour block format."
"What's one simple habit I can build this week to feel more productive?"

For Self-Check:

"Ask me 3 questions to help me evaluate my current season wisely."
"Help me build a rhythm of prayer, rest, and productivity."

For Emotional Strength:

"How can I encourage myself today without guilt or pressure?"
"Give me a grace-filled way to reset after I dropped the ball this week."

These prompts aren't about being weak. They're about being **strategic and supported.**

Final Word: You're Not Alone—And You Never Have to Build Alone Again

God gave you a calling. A team. A vision. But He also gave you wisdom to use tools that help you stay on track. AI, when used properly, becomes a **stabilizer** in your life. A tool that reminds you who you are, helps you focus on what matters, and moves you closer to what you're called to build.

Don't let your progress stall because you feel overwhelmed.

Don't stay stuck because you didn't know what question to ask.

Let the AI Life Coach serve you, so you can serve others. Let it help you reflect, realign, reset, and *move forward in peace.*

Because when you show up strong and stable, **your whole team benefits.**

Your success isn't just personal. It's shared.

So build yourself up—with wisdom, strategy, and Spirit-led support.

Then go build the Kingdom.

Chapter 20 – The Possibility of Getting More Done With Less Stress in Your Team Success

Exploring Gentle, Practical Ways AI Might Help You Better Manage Time, Tasks, and Energy—If You Need It

---

This May Help—But Only If You Need It

Not everyone feels overwhelmed. But many do. Not every Mutual Success Team is struggling with time management. But some are. And not every leader is burned out. But if you are... this chapter might be the fresh breeze you've been needing.

We're not here to sell AI as a cure-all. It's not. It's not going to make your life magically perfect. But if you've ever felt like you're spinning too many plates—or like your best intentions keep getting interrupted—this might help you get a better rhythm.

The key word here is **possibility.** This chapter is not a prescription. It's an invitation. A gentle one.

If you're curious, or if you've been looking for a little more ease in your day-to-day leadership, this might be a new way forward. AI might help. If not for everything, maybe just for that one task you keep pushing off. Maybe for that one meeting summary. That one list. That one message you keep forgetting to write.

You don't have to overhaul your system. Just explore a few ways to take back your peace.

Managing Tasks—One at a Time, With Less Mental Weight

Some people thrive with paper planners. Others feel like their brain is trying to hold twenty things at once—and it's tiring.

If you're in that second group, it may help to know that **AI can offload the mental weight.** Not by doing everything for you, but by helping you *see it all clearly*.

You might try:

"Help me organize my task list for the week by category—team projects, personal follow-ups, and outreach prep."
"Can you make this into a simple checklist with three top priorities per day?"
"Reformat this messy bullet list into a calendar I can follow easily."

AI can't tell you what to care about. But it can help *surface* what matters—so your energy goes to the right things, not just the loudest things.

If you already have a system that works—great. But if you've felt a little scattered lately, this might be worth exploring. You're still the leader. You still make the decisions. AI just helps you get there faster.

Managing Time—If You're Not a Natural Scheduler

Some people love calendars. Others... don't. Some teams are great at showing up on time. Others always feel behind.

If that's something your team struggles with—or if you personally feel like there's never enough time—it may help to ask AI for a little structure. Nothing rigid. Just some gentle scaffolding.

You could prompt:

"Break my 5 goals this week into time blocks that balance effort and rest."
"Help me build a 3-day plan to catch up on my missed tasks, without burnout."
"Show me a sample weekly rhythm for someone leading a Mutual Success Team and a small business."

Sometimes it's not about doing *less*. It's about doing *right-sized*. When time is reframed, pressure lifts.

Again—this isn't for everyone. But if you're spinning or stuck, AI can give you a starting point to breathe, reframe, and move forward with a bit more calm.

Managing Energy—For You or Your Team Members Who Feel Worn Out

This may not apply to you. But if you—or someone on your team—is running on empty, there's something sacred about simply giving yourself permission to **rethink the pace**.

You might explore prompts like:

"Help me plan a low-effort week that still honors my responsibilities."
"What are 3 gentle things I can do today that make tomorrow easier?"
"Write me an encouraging message to read when I feel behind or unmotivated."

Sometimes AI can serve like a non-judgmental companion—a sounding board that listens, sorts, and suggests. It doesn't push. It just points gently toward clarity.

If you already have rhythm, rest, and energy—that's beautiful. But if you've been wondering how to get your spark back, this might be a tool that gives you room to breathe.

Helping Your Team—Without Having to Be the Hero All the Time

If you're leading others, you already know: people need a lot. Support. Encouragement. Deadlines. Direction. And sometimes, it's a lot to carry.

If it feels like you're always answering questions, writing instructions, or organizing everything alone—AI might help *spread the weight.*

You could prompt:

"Write a training message for new volunteers who are helping with our Saturday outreach."
"Summarize this week's plan in a friendly message to the team."
"Create a checklist for someone helping us manage event setup."

This doesn't make you less valuable. It just helps you **show up as your best self—less drained, more clear, and more present.**

You're still leading. You're still guiding. But now, you're doing it with backup.

When It's Not a Fit—That's Okay Too

Some people may try all of this and feel, "It's just not for me." That's completely fine.

The heart of Team Success isn't tech—it's unity. It's progress. It's faith. And that's why this chapter doesn't give commands. It just opens a door.

For some of you, this door may lead to a season of **more clarity, less stress, and more joy** in your daily rhythm.

For others, you might close it—and that's okay too. You've got your flow, your systems, your Spirit-led schedule. And we honor that.

The power isn't in the AI. The power is in the permission: to explore, to release pressure, and to build a rhythm that works for you.

Final Word: There's Another Way—and You Get to Choose

You may not need to overhaul your life. You may not need another app. But if you've been wondering, "Is there a simpler way to manage what's on my plate?"—there just might be.

There's no pressure here. Just a gentle reminder that you don't have to do it all in your head. You don't have to carry every detail. You don't have to be the superhero every single day.

Whether it's organizing tasks, reclaiming time, or protecting your energy, AI may be one way you *lighten the load without losing the mission.*

Try it. Or don't. Either way, know this: you are seen. Your effort matters. Your leadership is sacred. And whether you use AI or not—you're still moving forward.

So here's to doing what works. With faith, peace, and freedom.

One step at a time.

Chapter 21 – Breaking the Scarcity Mindset: Talking With AI to Reinforce and Explore How You Can Overflow with Abundance

Using AI Conversations to Help You Renew Your Mind, Connect with Others, and Position Your Life for God's Overflow

---

There Is No Lack in the Kingdom of Heaven

If you're going to build a Mutual Success Team, you need to let go of the scarcity mindset. Not because it's trendy. But because it's *not Kingdom.*

Scarcity says, "There's not enough."
Abundance says, "There's more than enough—with God, through people, and together."

In Acts 4:34, it says *"There were no needy persons among them."* That's not poetic. That's a blueprint. That's what it looks like when believers trust God and one another. It's when independence becomes interdependence—and miracles happen in shared spaces.

But here's the tension: many Christians still live, think, and make decisions based on fear of lack. It's been drilled in from the world: "Protect your piece. Hide your gifts. Compete. Keep up. Do it alone."

AI can't fix your heart—but it can **help you renew your mind.** This chapter will show how simply "talking with AI" can become a way to challenge your limiting beliefs, reinforce Kingdom truth, and build the kind of connectedness that leads to true abundance.

When You Talk to AI About Lack, It Helps You Find Truth

Let's be clear: the Holy Spirit is your source of truth. The Bible is your foundation. But AI can be a sounding board—a tool to process what's swirling in your head and remind you what's possible.

If you've ever thought things like:

- "I'll never have enough."
- "People don't really help each other."
- "It's safer to go solo."
- "God wants me to struggle to stay humble."

You can bring those thoughts into the open—and ask AI to help reframe them.

For example:

"I feel stuck financially and like I'll always be behind. Can you help me reframe this in light of God's promises and Kingdom abundance?"

Or:

"Give me five truths that counter a scarcity mindset, written like a Christian life coach."

You don't have to "believe it" yet to begin the conversation. You just have to be open to the possibility of something more. AI becomes a coach that holds space for your renewal.

And as your mind shifts, so does your ability to receive.

Mutual Success Is the Engine of Abundance

It's hard to build abundance alone. You weren't meant to.

Scripture says we are one Body, many members. It says we are living stones, being built together. God designed you to *need others*—and to be needed.

The more we lean into Mutual Success—the kind of team where each person brings their best, gives what they have, and lifts up others—the more we see overflow.

AI can help here too. You might prompt:

"Help me create a Mutual Success Team vision document that invites others into shared abundance."
"Write a short message that explains to a friend why I want us to build something together—not just for income, but for Kingdom impact."
"What are five ways people in a small Christian group could each contribute a skill toward building a sustainable income project together?"

These conversations don't just produce answers—they shape *possibilities.* They get your mind working in collaboration mode. And when you combine that with faith? Things move.

Creating Abundance Through Strategic Unity

It's not just spiritual alignment—it's also practical wisdom. Here's how being part of a Mutual Success Team positions you for overflow:

1. Combined Skill Sets
You don't have to be good at everything. AI can help you list your gifts and match them with others.
Prompt:

"Help me inventory my strengths and list what kind of teammates would complement me."

2. Shared Resources
Instead of four people buying four tools, one person buys and shares—and AI helps you organize access.
Prompt:

"Create a fair tool-sharing plan for a small team using AI platforms."

3. Faster Decision-Making
Teams that prompt together move faster. One idea becomes four solutions.
Prompt:

"We're a group of believers with different business ideas. Help us design a system to vote and test quickly."

4. Encouragement and Faith-Building
When one person is low, the others lift up. AI can even help write those encouraging words.
Prompt:

"Write a message to a teammate who feels like giving up, reminding them of God's faithfulness and our shared vision."

5. Multiplication
Your small breakthrough gets duplicated across cities, churches, and teams. AI helps document and format it.
Prompt:

"Turn this testimony into a one-pager to inspire other churches to build a Mutual Success Team."

Get Into Position to Be Blessed

Abundance doesn't chase confusion. It chases clarity. It chases alignment. It flows through order, vision, and trust.

That's why simply *framing* your group as a Mutual Success Team creates a different atmosphere. It says: "This isn't random. This is sacred." It tells God, "We're ready to receive, to build, and to multiply what You place in our hands."

Even if you don't know exactly what you're building yet, the decision to build *together* positions you for provision.

AI can help with that positioning:

"Write a prayer to commit our team to shared purpose and open-handed abundance."

"Create a first meeting agenda for a Mutual Success Team forming to pursue overflow through shared Kingdom work."

These tools don't replace obedience—they help organize it. They help you do the *next right thing* faster, with more confidence.

Final Word: From Scarcity to Strategy to Overflow

You're not just here to survive. You're not just here to scrape by. You were made for Kingdom abundance—and part of walking in that is learning to **break the scarcity mindset** and build with others.

Talk with God. Talk with trusted teammates. And yes, talk with AI—if it helps—to process, renew, and plan.

You'll discover that abundance isn't just about money. It's about overflow in ideas, solutions, courage, and community. It's about becoming a fountain, not a container.

When you renew your mind, join with others, and ask for wisdom—you won't just get by. You'll start to build something that blesses you and everyone around you.

It starts with a mindset. It grows with a team. And it multiplies when faith meets action.

Chapter 22 – From Survival to Overflow: Using AI to Stabilize and Strengthen Struggling Christians

How Faith-Filled Conversations With AI Can Help You Regain Balance, Build Strength, and Move From Barely Making It to Thriving Again

---

It's Okay If You're Not Okay—But You Don't Have to Stay There

Not every season is easy. Some of us are building Mutual Success Teams from a place of strength. Others are starting while still struggling—emotionally, financially, mentally, or spiritually. And if that's you, this chapter is for you.

You may feel like you're in survival mode. Maybe you've been trying to hold everything together—your family, your business, your faith. It can feel heavy. Like one wrong move will cause the whole thing to collapse.

But here's the good news: God is still with you. And He's not just offering *relief*—He's offering *stability.* And in time, *overflow.*

While AI won't solve everything, it may serve as a quiet, supportive helper during these moments—especially when your strength is low and your clarity is foggy. It doesn't replace God, or church, or people—but it can help *stabilize* your day-to-day when things feel chaotic.

And from that place of stability, you begin to build again—with faith, with peace, and yes, with practical tools.

How AI Conversations Can Create Calm in a Storm

You don't need another complicated solution. You need something simple, kind, and steady. That's what certain AI tools can provide when used wisely.

They can help you:

- Organize your thoughts when your mind is racing
- Process hard emotions gently
- Write a prayer or reflection when you can't find the words
- Break your day into manageable steps
- Write encouragement to yourself—or someone else
- Reflect on what's going right, not just what's going wrong

For example, you might prompt:

"Help me reset after a discouraging week. I'm tired and don't want to quit, but I'm low on energy and hope."

Or:

"Write a 3-minute morning routine I can do to feel grounded, even if I'm overwhelmed."

The tone matters. If you approach AI like a taskmaster, it can feel cold. But if you frame it like a supportive coach—faith-centered, gentle, and honest—it becomes a companion on the journey.

Practical Ways to Stabilize Your Day (If You're Struggling)

Let's walk through some scenarios where AI may help you stabilize, *if you choose to use it.* These are not rules—they're gentle tools.

1. When You're Mentally Scattered

Prompt: "Help me organize today into three simple tasks that will help me feel stable and successful."
Follow-up: "Remind me why it's okay to do less when I'm rebuilding."

2. When You're Spiritually Dry

Prompt: "Give me a list of 5 Bible chapters to read when I feel like God is silent."
Follow-up: "Write a short prayer asking for restored joy and daily strength."

3. When Finances Are Tight

Prompt: "What are 3 ideas to earn extra income this week using basic skills and tools I already have?"
Follow-up: "Write a short budget plan that helps me feel hopeful—not ashamed."

4. When Relationships Feel Heavy

Prompt: "Help me write a message to someone I love but feel disconnected from. Keep it kind and faith-centered."
Follow-up: "What are 2 questions I can ask to start restoring connection?"

5. When You're Too Tired to Plan

Prompt: "Help me simplify my week into a gentle, faith-filled plan."
Follow-up: "Add one small thing I can do each day that reminds me I'm growing."

Each of these is a stepping stone—from survival to strength.

How This Helps Your Mutual Success Team Grow Stronger

If you're struggling, it's tempting to withdraw. To stay silent. To push through quietly, hoping no one notices. But in a Mutual Success Team, your presence matters—even in your weakness.

When you allow yourself to receive support—spiritual, emotional, and yes, even digital—you're modeling something powerful. You're showing that strength isn't about pretending. It's about *getting help that works.*

And as you stabilize:

- You show up more consistently
- You contribute your unique gifts
- You offer empathy to others who are hurting
- You become part of the overflow—not just a recipient of it

AI, used wisely, can help you regain footing faster. It's not the source of your breakthrough—but it can support the process.

Prompt:

"Write a note I can share with my team, letting them know I'm getting back on my feet and still committed."

That's what builds trust. That's what builds teams that last.

Overflow Doesn't Come From Trying Harder—It Comes From Rebuilding With Peace

Your goal isn't just to function. It's to *flourish.* But flourishing doesn't come from hustle. It comes from rootedness. And rootedness takes time, space, and support.

AI may not be your solution—but it may become your structure.

A few supportive prompts, some encouraging words, a well-timed plan... and suddenly, the fog lifts a little. You feel grounded. Seen. Capable.

Overflow will come. Not all at once, but steadily. As you recover your clarity, as you organize your days, as you rebuild your rhythms—you'll start to notice your capacity increasing.

And when your cup begins to overflow, you'll be ready to pour into others—not from pressure, but from fullness.

Final Word: Survival Isn't the End of Your Story

God's plan for you doesn't stop at survival. He wants to bring you to *overflow.*

And if right now, you're somewhere in between—AI may be one tool you use to take that next quiet step. Not to escape hardship, but to walk through it with grace, clarity, and support.

You're not alone. You're not too far gone. You're not disqualified.

You are in a season. And seasons shift.

Use what helps. Let go of what doesn't. And keep moving forward—faithfully, gently, steadily.

Let your Mutual Success Team support you. Let the tools God has made available work for you. And trust: overflow is not just possible.

It's promised.

Part 5: Tools – Breakthroughs Can Happen When You Use Several AIs Together

This part dives deep into one of the most practical revelations in the book: using more than one AI tool at a time. Instead of limiting yourself to one system, readers are shown how each major AI platform offers something slightly different—creating the opportunity to multiply insights and find better answers.

Some tools offer greater creativity, others stronger research, while others still provide a more human-sounding response. Instead of depending on just one source, this part teaches readers to prompt across platforms, compare outcomes, and gather wisdom from multiple voices. It's not about information overload—it's about refining clarity.

With simple systems like Google Docs, paired with cross-platform prompting, "Mutual Success Teams" can build tools, write flyers, brainstorm ideas, and organize massive resources in a short amount of time. When each team member learns this skill, the whole network gains momentum that would be impossible to create manually.

The main takeaway here is this: don't settle. There is wisdom in a multitude of counselors, and in this case, those counselors are free and always available. Learn the strengths of each AI, and watch how using them together helps your "Team Success" projects rise faster and stronger than ever before.

Chapter 23 – Miracles in the Making: Using AI to Help Find Real Answers and Real People of Faith

How to Tap Into the Profound, Life-Transforming Possibilities of AI by Asking Real Questions—And Seeking Real Miracles

---

AI Isn't Just a Chatbot—It's a Tool for Real, Practical Miracles

Many people are still using AI like it's just a novelty. A clever chatbot. A helper for simple things. But what if we told you that **AI has the potential to unlock real breakthroughs—miraculous ones—if you just take it seriously enough to ask better questions?**

That's the key. It's not that AI is "miraculous" on its own. It's not supernatural. It doesn't carry the Spirit of God. But it is a *practical tool*—an incredibly powerful one—that sits idle until someone with faith, vision, and boldness picks it up and *asks it to do more.*

The Body of Christ is suffering in too many ways—confusion, lack, sickness, isolation. And now that we have something this effective, this scalable, and this available? **We are without excuse.** We must start using it with purpose.

We must stop thinking "chatbot" and start thinking "miracle assistant." We must stop asking surface questions and start asking for profound, transformative results. It's time to raise our expectations.

Real Miracles Begin With Real Questions

The reason most people don't get life-changing help from AI is because they don't ask for it. They skim the surface. They stay polite, safe, or shallow. But when you bring real hunger to the conversation—AI becomes something else entirely.

One example from this author's life proves this. In the book *Faith Healing Trainings*, he described a conversation with AI that changed everything. It began not with a clever prompt, but with a metaphor from the heart.

He said:

"I want to understand healing and faith. So let me tell you this story. At church, there's a chair that flips up to save space. But my friend knows how to set a hidden latch that keeps it down. I want to see the faith process like I could turn that chair upside down, find the latch, and understand how it works."

And something happened.

The AI didn't just give generic Bible verses. It didn't just give summaries. It gave *clear insight*—in metaphor, structure, and wisdom. Because the question was deep, raw, and hungry.

When you bring your real self to AI—desperation, faith, emotion, creativity—you start to experience something different.

You don't get miracles by being casual. You get them by being *committed.*

Use AI to Discover People Who Can Help You Receive Your Miracle

AI isn't just about generating content—it's also about searching out the people, teachings, and resources God may use to deliver your healing, wisdom, or breakthrough.

In this author's journey, AI helped surface something most believers still don't know exists: the **Divine Healing Technician Training (DHT)** by Curry Blake. Free on YouTube. Biblically sound. Powerfully effective. No-nonsense. No fluff. And with results.

Curry teaches that God *always* wants to heal. That you, as a believer, have the authority. That you don't need to beg God—you need to believe Him. And hundreds of thousands have received healing as a result.

AI helped find it. Not by accident—but by asking:

"Who is teaching guaranteed healing results in Jesus' Name today?"
"Is there someone helping people renew their minds for miraculous healing?"
"Show me the strongest free teachings on Divine Healing that work."

The answers came. Links. Names. Channels. Conferences.

And now that tool—AI—is part of how the Body of Christ connects to the Body itself. It becomes a **searchlight into the hidden parts of God's provision.**

You Don't Have to Stay Stuck—Ask More Real Questions

Many people never get to the depth of what's possible because they never *really* ask.

They ask:

"Can you write a prayer for healing?"
But never:
"Help me build a daily faith training routine that reprograms my thinking for Divine Health."

They ask:

"Who in the Bible was healed?"
But never:
"Explain the difference between believing in healing and *believing for healing* with precise examples."

They ask:

"What is faith?"
But never:
"Write a 5-day faith-building challenge that strengthens my belief with exercises, scriptures, and metaphor."

You don't need perfection—you need *hunger.*

When you ask serious questions, AI responds in serious ways. It's not magic. It's about using the tool to reach beyond where you've been.

You are not waiting for your miracle anymore. **You are part of making it.**

Miracles Come Through the Body—And AI Helps You Find the Rest of Yours

You were never meant to walk alone. And miracles—whether healing, financial provision, deliverance, or breakthrough—rarely come in isolation. They come through *people.* Through unity. Through those who are walking in a grace you haven't yet accessed.

AI can help you find those people.

Prompts you can try:

"List current faith-based ministries where healing is taught and regularly confirmed."
"Who is teaching radical faith for abundance and physical healing, with a practical daily system?"
"Find resources that can help me remove doubt and strengthen belief in God's promises."

Then ask for summaries. Key ideas. Action steps. Start learning. Start applying.

You may not get your miracle from AI. But you might get the *path* that leads you to it.

And that alone... is a miracle in the making.

Final Word: God Has Made the Tools Available—Now It's Time to Use Them

You already have what you need:

- A Spirit-led heart
- A real hunger
- A powerful team
- And now... a tool that listens, thinks, remembers, and responds with clarity 24/7.

AI isn't the answer. But it can lead you to the answer.

And when used with intention, boldness, and faith—it becomes a magnifier of truth, a finder of resources, and a builder of breakthrough.

So go ahead. Ask the deeper questions. Seek out the hidden teachers. Design your own miracle-training pathway. Invite God into the conversation. Invite your Mutual Success Team into the learning.

And get ready.

Because your miracle may be one real question away.

Chapter 24 – Discovering Real World Help: Using AI to Find Ministries, Healers, and Resources for Mutual Success Needs

How to Search Smarter Across All AI Platforms to Find the Real Answers and Real Support You Need—So No One in Your Team Goes Without

---

When You're Searching for Help, Make the Search Count

One of the most underused features of modern AI tools is this: **search.** Not just casual searching. Not just "who won the game last night" kind of searching. But real, intentional searching—on behalf of your Mutual Success Team, your family, your health, your spiritual life, or your business.

And here's the truth: If you only search once, on one AI platform, you're likely missing something big.

AI tools are constantly changing, constantly learning, and they all pull from different sources, with different strengths. So when you're trying to find real help—whether it's a local Christian counselor, a healing ministry, a faith-based housing program, or someone who teaches Divine Healing—you can't afford to stop early.

This chapter is your guide to using **all four major AI tools**—ChatGPT, Claude, Gemini, and Grok—as a united search team. It's your reminder that God wants to meet your needs, and that now more than ever, you have tools in your hand that can uncover real-world help, fast.

Why Searching on One AI Isn't Enough

Each AI platform has a unique way of seeing the world. They access different data sets. They summarize information differently. And they apply tone and priorities in different ways.

That's why, when you run the **same prompt on four tools**, you often get four completely different answers. One might point you to a YouTube channel you've never heard of. Another gives a name and contact info of a local ministry. A third gives you a list of principles, and a fourth gives you a full action plan.

If you only try one? You might miss your answer entirely.

The author of this book learned this the hard way—until he started **copying and pasting his prompts into all four tools**. He'd write out his real need, like:

"I'm a Christian looking for healing resources that actually produce results. Are there ministries or teachers with free online training that works?"

By running this prompt on all four platforms, he found answers he *never* would have gotten from a single tool. Curry Blake. DHT. The "New Man" conference. Resources on healing, belief, authority, and faith. Some came from ChatGPT. Others from Gemini. And still more from Claude.

You're not using AI the right way unless you're searching wide. You're not stewarding the tools unless you use *all of them.*

How to Use All Four AI Tools for Maximum Discovery

Let's get practical. Here's a 3-step process you can use anytime your team is looking for help—whether spiritual, emotional, financial, or physical.

Step 1: Create a Search Prompt With Detail and Emotion
Don't be vague. Be clear and heartfelt.
Example:

"We're a Christian team in Texas trying to launch a healing outreach. We're looking for ministries, resources, or leaders who could train us or guide us. Ideally free. What's available?"

Step 2: Copy and Paste That Prompt Into All Four Tools

- **ChatGPT** (chat.openai.com)
- **Claude AI** (claude.ai)
- **Google Gemini** (gemini.google.com)
- **Grok** (via X.com or built into X Pro)

Paste the prompt into each AI. Don't change it. Let each one give its unique perspective.

Step 3: Compare, Collect, Act

- Write down the best recommendations
- Click the links, watch the videos, read the summaries
- Ask follow-up questions on anything unclear
- Organize results into a Google Doc for your team

Now you're searching like someone who expects results—not just information.

What Types of Needs Can Be Met Through AI Search?

This is where it gets powerful. You can use AI to find real-world help for almost *any* area of struggle or need in the Body of Christ. Here are just a few:

Healing & Health

"Find faith-based healing ministries with documented testimonies and free trainings."
"Are there Christians teaching physical healing through prayer with results?"

Housing & Recovery

"What Christian ministries offer transitional housing in [your city]?"
"Are there local programs that help struggling single moms in the church?"

Financial Provision & Work

"List Christian business programs or side-hustle coaching opportunities aligned with faith."
"Any faith-based job training or remote work programs for believers?"

Mental & Emotional Support

"Find Christian counselors or support ministries online that offer sliding scale or free care."
"Are there pastoral care groups for Christians dealing with anxiety?"

Teaching, Training, Equipping

"Give me a list of Christian conferences or online summits focused on healing, deliverance, or the gifts of the Spirit."

All of this can be uncovered in less than an hour—**if you search well.**

God Wants to Meet Your Need—Don't Stop Searching Until It's Met

You don't have to live in lack. You don't have to stay stuck. And you certainly don't have to do it all alone.

God's design is that there would be *no need among them*. That means you. That means your Mutual Success Team. That means every person God has called you to serve.

But we also have a role to play: **Search. Knock. Ask. Seek.**

Don't give up after one AI doesn't help. Try the next. Rephrase the prompt. Ask from a different angle. You wouldn't stop praying because your first words didn't bring an answer. You keep asking. You press in.

In the same way, AI tools respond to persistence. Keep looking. Keep prompting. You'll be shocked at what is out there—waiting for the Body of Christ to pick it up and run.

Final Word: Use the Tools, Find the Help, Meet the Need

You are equipped for overflow. And overflow doesn't mean doing it all yourself—it means using *everything God placed in your hands* to meet the needs around you.

AI search isn't fluff. It's not extra. It's part of how we now access real people, real places, and real help.

Start now. Write your need clearly. Run your search wide. Build a resource library. Share it with your team. Organize it for others.

And in doing so, you become not just a receiver—but a provider.

You'll be amazed at what's been hidden in plain sight.

Chapter 25 – The Phone vs The Computer: Choosing the Best Tools for Deep Mutual Success Development

Understanding the Strengths and Limitations of Each Device—And Why Using Both Together Can Multiply Your Team Success Results

---

The Right Tools Make the Mission Easier

There's a reason Jesus said, *"Count the cost."* When you're building something lasting—like a Mutual Success Team—you want to make sure you're using the best tools available to you. Not just spiritually, but practically.

In today's world, the two most important digital tools are the **smartphone** and the **laptop computer**. Both are powerful. Both are capable of helping you run a business, grow your outreach, and train your team. But they're very different. And understanding the strengths and limitations of each is the key to maximizing what God wants to do through you.

This chapter is about getting strategic—not just with how you use AI, but *where* you use it. Because the truth is, you can do a lot with just your phone. But when you combine it with the power of a laptop? That's when you start building *deep development* for your Mutual Success.

It's like the old Captain Planet line: *"By your powers combined!"* For us, it's: *"By these two tools combined—we can build something great."*

Why Phones Are Great (And Where They Fall Short)

Let's start with the tool most people have: the phone. It's convenient, portable, and almost always in your hand. You can:

- Run all four major AI tools as apps
- Prompt AI for help on-the-go
- Take quick notes or screenshots
- Use voice-to-text to brainstorm
- Send team messages instantly
- Record video or audio reflections

Phones are powerful for capturing ideas, solving small problems quickly, and staying connected. When you're at a grocery store, waiting in a line, or lying in bed—you can still lead, create, and plan.

But here's where phones fall short:

- Small screen = limited vision
- Typing is slower, formatting is harder
- Copying/pasting across apps can be clunky
- It's harder to organize or see full documents
- Some tools just don't work as well in mobile apps

- You can't multitask effectively or drag/drop files

That's not a deal-breaker. But if you *only* use your phone, you're likely working 50% slower—and with more stress.

Why Computers Unlock a Whole New Level of Productivity

A basic laptop doesn't have to cost $2,000. Many solid ones range between **$500–$700** and can last 5+ years with care. When you start working on a computer—even a modest one—things shift. You can:

- Open multiple AI chats in different tabs
- Copy and paste prompts and answers quickly
- Save and organize documents on your hard drive
- Use Word, Google Docs, or Free Office with full formatting
- Manage complex projects with spreadsheets and folders
- Edit graphics in programs like Canva or Photoshop
- View your work on a full 15-inch screen, clearly

This is where the deeper "development" of a Mutual Success Team becomes possible.

You're not just reacting. You're building systems. Templates. Courses. Documents. Plans. These are the kinds of tools that **turn your ideas into duplication**—so other churches and teams can learn from what you're doing.

Think of your laptop as your **studio.** It's where you build what will bless others.

Using Both Together: A Strategy for Serious Builders

You don't have to choose one or the other. The real power comes when you **use both devices together**, as a tag team.

Here's how that might look in real life:

- Use your **phone** to brainstorm prompts or record a voice note with your idea
- Later, use your **computer** to format that idea into a teaching, flyer, or document
- Ask AI a deep question on your **phone** while driving or walking
- Then copy that result into your **laptop**, clean it up, and save it to your training files
- Run a **team Zoom** call from your computer, while referencing your notes from AI chats on your phone
- Use the **phone camera** to record a testimony, and the **computer** to edit and upload it professionally

This is not complicated. It's just intentional.

When you make each device play to its strengths, your productivity skyrockets. You're no longer wasting time or losing ideas. You're building momentum.

What If You Don't Have a Computer Yet?

That's okay. You're not behind. But it may be time to *consider investing in one*. Because the kind of leadership, system-building, and documentation that creates long-term Mutual Success will eventually require more screen space, more power, and more organization.

You don't need the fanciest setup. But here's what to look for:

- 15-inch screen (for clear visibility)
- 8GB RAM or more (for multitasking)
- Keyboard you enjoy typing on
- $500–$700 price range (new or refurbished)
- Can run modern browsers and office apps easily

Then start using it to build:

- Mutual Success Playbooks
- Team Training Curriculums
- Templates for documents, flyers, and emails
- Testimony folders
- Project summaries
- Income systems for other churches to model

This is where you multiply—not just produce.

You go from surviving... to building legacy.

Final Word: Use What You Have—and Plan for What's Next

You don't need everything today. But you do need a strategy.

If all you have is a phone—use it well. Learn to prompt. Capture your thoughts. Search smarter. Move forward.

But if you're serious about building something that lasts? **Make it your goal to add a laptop to your toolset.** It's not a luxury. It's a leverage tool. And when paired with your phone, it becomes the "combined power" that makes your vision possible.

We are called to build wisely. Faithfully. Skillfully.

And with these two tools—AI on your phone and on your computer—you'll be equipped to lead, train, bless, and multiply like never before.

This is what deep development looks like.

You're not just using devices. You're using your God-given mind, heart, and resources to build the Kingdom—with excellence.

Let's build it well. Together.

Chapter 26 – Google Docs + AI: Organizing Your Mutual Success Team Knowledge for Long-Term Impact

How Combining These Two Free Tools Can Unlock 20 Powerful Benefits for Your Personal and Team Growth

---

When You Combine the Tools, You Multiply the Results

If you want to build something that lasts, you don't just need big ideas—you need a **system for organizing those ideas**. The most powerful breakthroughs often never multiply because they never get written down. And even when they are, they get lost in notes, emails, or scraps of paper.

That's where the combination of **AI tools + Google Docs** becomes one of the most powerful, free, and scalable systems you can use as a Christian leader or Mutual Success Team builder.

On their own, each tool is helpful. AI gives you speed, insight, structure, and support. Google Docs gives you access, collaboration, and storage. But when you use them *together*—intentionally—they become a **platform for building your legacy**, one page at a time.

Whether you're creating training guides, project plans, testimonies, business models, or team playbooks, this combo becomes your **digital archive of Kingdom progress**. You're not just working—you're building something others can follow.

Let's break down how this works, and explore 20 unique benefits—10 for your personal success, and 10 for business growth with your Mutual Success Team.

---

10 Benefits for Personal Success When You Combine AI + Google Docs

These benefits are especially for you—the builder, the believer, the person who's organizing life, faith, and work day by day.

1. You Never Lose a Great Idea Again
When AI helps you brainstorm or outline something amazing, you can instantly save it in a Google Doc and revisit it anytime.

2. Your Thoughts Become Organized and Searchable
Instead of 500 notes in your phone, you have clearly labeled documents—topics, titles, and links—ready to go when needed.

3. You Can Reflect on What God Has Shown You
Record your insights, prompt responses, prayers, and revelations into folders by date or topic. You now have a digital spiritual journal.

4. You Reduce Decision Fatigue
Use AI to help clarify priorities, then store your top 3–5 weekly goals in one doc you can check each morning.

5. You Can Plan Ahead With Peace
Use AI to generate a 30-day personal plan, save it in Google Docs, and check in weekly for course corrections.

6. You Can Build a Personal Library of Encouragement
Save AI-written devotionals, affirmations, or encouragements under topics like "when I feel tired" or "when I need courage."

7. You Save Time
No more writing things from scratch every week. Store, duplicate, and reuse what works—quickly.

8. You Train Yourself for Excellence
Each time you ask AI to improve something you've written, and save the results—you build a better "you" library.

9. You Can Access Your Documents From Anywhere
Phone, tablet, laptop, library—you're not stuck. Your tools are with you, always.

10. You Create a Track Record of Growth
Months from now, you'll look back and see the evidence of what God has built through your consistency—and it's all documented.

---

10 Benefits for Business & "Mutual Success Team" Growth Using AI + Google Docs

Now let's shift the lens to what happens when you use these two tools not just for yourself—but for your whole team or your growing business.

1. You Create a Team Playbook
Every insight, policy, win, and idea can become part of a living team manual that grows with you.

2. You Make Training Repeatable
Once you train one person, you save the process in a document. Now you can train 100—without extra work.

3. You Multiply Your Projects
Successful events, launches, outreaches? Document the plan with AI, save it, and replicate it in new places or churches.

4. You Onboard People Faster
Use AI to create a new member welcome doc. Save it in your Team Drive. Share it with one click.

5. You Run Smoother Meetings
Create a shared team agenda. Let AI help you summarize it. Store it where everyone can edit and see it live.

6. You Track Team Progress Publicly
Create a shared Google Sheet of weekly action items. Use AI to automate follow-up summaries and accountability.

7. You Delegate with Clarity
Use AI to create role descriptions. Store each one in a folder with to-do checklists and documents linked.

8. You Build a Resource Library for Others
Every flyer, testimony, video script, process, or template you create can be used by churches or teams around the world.

9. You Stay Aligned as a Team
Document your shared vision, values, priorities, and strategies. Review them monthly. Let AI help refine them over time.

10. You Create Something Worth Handing Off
Whether you're training up new leaders, or one day passing the baton, your knowledge lives on—accessible, clear, and ready.

Start Simple—Then Let It Grow

You don't have to build a 100-document archive tomorrow. Just start with one. Let AI help you write your team's weekly goal list. Save it in a doc. Then next week, do it again. Add a second doc for outreach ideas. A third for testimonies. A fourth for resources.

In time, you'll look back and realize—**you've built a Mutual Success Library.**

People will ask, "How did your team grow so fast?"
And you'll know: it wasn't just meetings. It was documentation.

You used tools well. You captured what worked. You created clarity where others stayed vague.

This is how movements grow.

Final Word: Don't Just Work—Build Something That Lasts

You are not just a dreamer. You are a builder. And builders use tools.

Google Docs + AI are two of the most powerful tools you'll ever use. One is your thought partner. The other is your archive. Together, they help you **turn God-given vision into God-honoring systems.**

So don't let your best ideas fade. Don't let team lessons get lost. Don't waste time repeating what could've been saved.

Instead—build the system now. So others can run with it later.

You're not just organizing documents. You're organizing destiny.

Let's make it easy to grow, easy to train, and impossible to forget.

Part 6: Training The "Mutual Success Team" – Reliable, Fast, Customized

Equipping the team is no longer slow or complicated. This part of the book outlines how AI enables fast, custom-built training for any member of a "Mutual Success Team." Whether onboarding someone new or sharpening a veteran leader's skillset, AI can generate exactly what's needed in minutes.

The process is revolutionary. Ask AI to write a course outline, fill it with detailed steps, and format it into shareable documents. Team leaders can now train others without draining time, and members can get personal mentorship and instruction whenever they need it. There's no longer a bottleneck.

This isn't just about convenience. It's about kingdom impact. Imagine training 50 people at once—each with material tailored to their understanding level, learning style, or spiritual maturity. That's what this part makes possible. Customization becomes the normal way of equipping every person for their unique role in "Team Success."

By the end of this section, the reader is empowered with repeatable methods for passing on wisdom, documenting lessons, and building duplicatable trainings that scale. The goal is not complexity—it's clarity, simplicity, and momentum. Training has never been this accessible or this Spirit-led.

Chapter 27 – Saving Documents For the Long-Term: Using Google Docs to Save & Share

How to Capture, Organize, and Distribute What You Learn—So Others Can Be Strengthened, and Mutual Success Can Multiply Across the Globe

---

If You Don't Save It, You Might Lose It

Have you ever had a moment of breakthrough—an insight, a solution, or a training idea—and thought, "I'll remember this later"? And then... you didn't?

That's not forgetfulness. That's being human.

And that's why this chapter is so important.

If we're going to build churches, ministries, businesses, and Mutual Success Teams that actually create *lasting* impact, we must become **people who document what we're learning.** Because one lesson saved is worth ten repeated. One insight recorded can bless not just you—but *hundreds of others.*

And now, thanks to **Google Docs**, there's a free, simple, shareable way to save everything that matters—from revelations to checklists, from testimonies to how-to guides. You're not just creating files. You're creating *legacy.*

This chapter will show you why it matters, how to do it practically, and how your saved documents can help expand the mission of the Team Success Network—so "there is no need among them" becomes a reality, not just a dream.

Why Long-Term Saving Makes Short-Term Growth Easier

Think of your life like a book. You don't just live it—you write it. And if you want others to learn from what God's done in your life or through your team, you have to **put the story on paper.**

When you use Google Docs intentionally, you:

- Remember what worked
- Avoid repeating mistakes
- Build upon past wins
- Empower others without starting from scratch
- Create team unity around documented truths

Let's say your team runs an outreach that goes better than expected. You used AI to help plan it, you had a great turnout, and you learned 3 lessons you want to apply next time. If you don't save it, you risk forgetting key steps.

But if you immediately open Google Docs and create:

- "Outreach Checklist – What Worked"
- "Post-Event Team Reflections"

- "Flyer Templates + Invite Language"

Now, that one event becomes a **repeatable system.** And you can share it with your church, another team, or with Team Success Network.

What to Save: Not Everything—But the Right Things

Here's a powerful secret: You don't have to save *everything.* You just have to save what multiplies.

Here are some of the most valuable things to capture in Google Docs:

1. Mutual Success Insights
Any moment where something "clicked" for your team—record it. What was the insight? How did it happen? How might others use it?

2. AI Conversations With Value
When an AI gives you a powerful answer, copy/paste it into a Doc. Label it clearly: "Healing Advice," "Leadership Flowchart," "Conflict Resolution with Love."

3. Weekly Team Plans
Keep running logs of your team's goals, action items, and summaries. You'll start to see growth patterns over time.

4. Checklists That Worked
If you built a great meeting agenda, onboarding process, or event setup list—don't lose it. Save it in a shared folder.

5. Training Scripts or Devotionals
Did you give a short talk or encouragement that resonated? Save it. That content may train someone across the world one day.

6. Testimonies
Any time someone on your team has a win, record it. Even short bullet points are fine. These are seeds of encouragement.

7. Lessons From Mistakes
Failure is the best teacher—if you write it down. What didn't go well? What will you do differently? Future-you will thank you.

8. Your Own Reflections
Take 5 minutes a week to write: "What did I learn this week about Team Success?" Over time, these become powerful summaries of wisdom.

How to Share What You Save With the World

Saving is just step one. **Sharing** is where the Kingdom multiplies.

Thanks to Google Docs, it's simple. One click gives anyone access to what you've built. You can:

- Share with your team
- Share with your church
- Share with your new members
- Share with the Team Success Network
- Share with another Christian across the world asking the same question you once had

Imagine hundreds of teams sharing their best documents—AI insights, plans, flyers, strategies, prayers, success stories—all collected in one place.

We're not there yet. But that's the future we're building toward.

You can send your saved docs directly to **results@TeamSuccessNetwork.com**, and your insight may help expand our global training library for churches, ministries, and individuals worldwide.

You're not just saving for yourself. You're saving for the *Body.*

Final Word: Teach What You Learn—And Build the Kingdom, One Doc at a Time

When God teaches you something that works—*save it.*
When you lead something that succeeds—*document it.*
When AI helps you find a shortcut, a breakthrough, or a blessing—*copy it into a doc, give it a title, and share it.*

This is how we grow.
This is how we teach.
This is how we build discipleship movements that are scalable, duplicatable, and long-lasting.

It doesn't take a fancy system. It takes commitment.
You're building something worth remembering. So make it savable. Make it shareable. Make it usable for someone else.

Let your journey become someone else's shortcut.
Let your team's breakthrough become a playbook for others.
Let your Mutual Success become our **shared success.**

Because in the Kingdom, what you save—can save others time, stress, money, and even faith.

And that's the kind of stewardship that blesses generations.

Chapter 28 – Creating Kingdom Courses & Learning Curriculum: Letting AI Format Teachings Into Courses for the Church

How to Turn Your Insights Into Full Courses With AI, So You Can Teach, Train, and Transform Others in Faith and Business

---

You're Already Learning—Now Multiply It Into Something Lasting

Every time you talk to AI, you're learning something. You might be solving a problem, exploring a verse, or asking about a strategy. But at some point, you'll feel it—that hunger to go *deeper*. You'll want more than a quick answer. You'll want a *framework*. A system. A learning journey.

And that's when it's time to turn your learning into a **course.**

This chapter is your permission to stop thinking of yourself as "just a learner" and to start seeing yourself as a **curriculum creator**. Whether it's for your own growth or for your team, church, business, or Mutual Success project—**AI can help you format your thoughts into full, structured, Spirit-led training.**

And once that training is saved in a Google Doc, you've created something powerful: a **Kingdom resource**. Something you can teach. Share. Multiply. Improve. Translate. Use again.

You don't need to wait for someone else to write the course you're craving. You can build it. You're qualified to begin. And AI is ready to help.

How to Ask AI to Create a Course for You

Creating a course with AI starts with one prompt. Something like:

"Help me create a full 4-week training course on how a Christian can become consistent in their faith, habits, and prayer life."

Or:

"I want to create a course for small Christian business teams on launching their first income-generating project using AI. Give me a full outline."

In response, you'll get:

- A **course title**
- A **description** of who it's for
- A **lesson-by-lesson outline**
- A set of **goals and objectives**
- Ideas for **interactive exercises**
- **Homework** or implementation steps
- Even **quiz questions**, if you want them

From there, you can say:

"Great. Now write the first lesson in detail."
"Can you include scriptures and discussion questions?"
"Add an opening story to make this lesson engaging."

In a single afternoon, you can co-create a full **discipleship journey**, a **training workshop**, or a **video curriculum**—custom-made for your audience, your church, or your calling.

And once it's done? **Save it in Google Docs.** Share it. Teach it. Tweak it. Repeat it.

---

10 Courses You Can Create for Your Own Personal Life Success

These ideas are about growing you. About becoming the kind of person who lives in wisdom, overflow, and purpose.

1. **Breaking Limiting Beliefs With God's Word**
2. **Building a Daily Routine That Aligns With Your Calling**
3. **Faith-Filled Finances: Budgeting and Believing at the Same Time**
4. **Becoming a Kingdom Communicator (Speaking With Power and Grace)**
5. **Hearing God's Voice: Developing a Personal Practice of Discernment**
6. **Healing From the Past: A Journey Through Forgiveness and Freedom**
7. **Radical Gratitude: Living from Joy in Every Season**
8. **Planning Your Week With the Holy Spirit**
9. **Faith Habits That Last: 30 Days to Transformation**

10. **Personal Calling Clarity: Discovering and Describing Your God-Given Assignment**

Each one of these can be prompted, formatted, and built with AI. You bring the heart. AI helps build the structure.

---

10 Courses You Can Create for Business Success With Your Mutual Success Team

These ideas are about equipping your whole team to grow something profitable, duplicatable, and rooted in Kingdom excellence.

1. **Launching Your First Mutual Success Project (Step-by-Step)**
2. **How to Use AI to Build a Faith-Based Business Plan**
3. **Fundraising for Good: How to Raise Support Without Fear or Pressure**
4. **Team Communication: Keeping Everyone on the Same Page**
5. **Kingdom Marketing 101: Messaging That Attracts and Honors God**
6. **Writing Powerful Testimonies to Build Trust and Multiply Impact**
7. **Documenting Your Systems So Others Can Succeed Too**
8. **How to Host a "Start Your Mutual Success Project" Workshop**
9. **Managing Conflict With Love and Clarity on Your Team**
10. **Training Others to Train Others: The Path to Duplication**

Imagine creating even just two or three of these. Imagine sharing them. Imagine watching other teams succeed because you were willing to **turn what you know into something teachable.**

You Don't Need to Be a Professional—You Just Need to Be Willing

The greatest courses are not always made by polished speakers or famous names. They're made by people who care. People who say, "This helped me—and I think it might help you, too."

With AI and Google Docs, you have what you need to start right now:

- You don't need a platform. Just a prompt.
- You don't need a publisher. Just a Google account.
- You don't need to be an expert. You just need to *care enough to start.*

Let AI guide your structure. Let the Holy Spirit guide your heart. And let your own learning be the seed that grows into someone else's success.

This is what discipleship looks like in the digital age.

Final Word: Teach What You Know—And Let the Church Grow

Don't let the learning stop with you. Don't hoard what God has given. Multiply it.

You can create a course. You can teach what you're learning. You can build a curriculum that will outlast your meetings, your season, and even your team.

So ask AI today:

"Help me create a course on __________."

Fill in the blank. See what happens. Then open Google Docs, create a new file, and paste it in.

You just started something that could change lives.

And when you're ready—**share your course with the Team Success Network.** Together, we're building a movement where Kingdom truth is accessible, teachable, and scalable.

One lesson at a time.

Chapter 29 – Custom Courses, Instant Training: Using AI to Equip New Team Success Members in Minutes

Why Personalized, Fast, and Spirit-Aligned Training—Created With AI—Is the Most Practical and Powerful Development Tool We've Ever Had

---

Training Just Changed Forever

In the past, onboarding someone to a new team, church department, or Kingdom project could take days, weeks, or even months. You needed someone with time. With experience. With teaching skill. And with a lot of patience.

But now? **Everything has changed.**
AI allows you to create **custom training—on demand—for any new person, any role, any project, any time.** And the results are nothing short of revolutionary.

You no longer have to say, "I don't have time to train someone." Because now, your AI assistant can do most of the work for you. You don't have to say, "I wish I had a training program for this." Because now, you can create one in **under 15 minutes**.

And best of all? It's **customized.** Tailored to your team's mission, your language, your style. This isn't cookie-cutter. It's made for you—and by you—with the help of AI.

Whether you're onboarding someone into a Mutual Success Team, assigning them a new responsibility, or helping them learn a concept quickly, you're no longer stuck. You now have a solution that is fast, free, and deeply transformational.

How It Works: From Idea to Curriculum in Minutes

Let's say someone just joined your Mutual Success Team. Their background is different. Their learning style is unique. And they're eager—but overwhelmed.

What do you do?

Simple. You ask AI:

"Create a 5-day onboarding training plan for a new Mutual Success Team member. Make it faith-based, encouraging, and include scripture."

In 30 seconds, you have:

- An outline
- Daily focus topics
- Scripture references
- Reflection questions
- Action steps
- Encouragement built-in

You then follow up with:

"Now write the full text of Day 1 in a friendly tone."
"Add three bullet points on how this connects to our team's current goal."
"Summarize it as a short email or printed handout."

Then, open Google Docs. Paste it in. Save it. Share the doc with your new team member.

That's it.
You didn't need a curriculum department. You didn't need an extra staff person.
You just created instant training—at the exact moment it was needed.

Why This Is So Profoundly Transformational

Here are just a few reasons why this method is more than just "cool." It's deeply spiritual, extremely strategic, and exactly what the Church needs right now.

1. Everyone Learns Differently
Now you can say, "Create a visual-based summary," or "Create a story-based version of this training," or "Write this as a conversation between two believers."
You're no longer stuck with a one-size-fits-all training program. AI lets you **adapt the message to the learner.**

2. You Can Train In Their Language
Literally. Ask AI to translate your training into Spanish, French, Swahili, or any major language. Or adapt it into simpler language for someone who struggles with reading. **No more barriers.**

3. You Can Update Training Instantly
If your team strategy shifts? Just ask AI to revise the document. No reprinting. No re-recording. Just adapt and go.

4. You Empower New Leaders Faster
Instead of waiting months to train someone to train others, you can hand them AI-generated tools and say, "You've got this. Let's go."

5. You Increase Retention and Ownership
When people feel supported with clear, customized training—they stick around. They show up. They grow.

And that's just the beginning.

Dozens of Other Amazing Benefits of AI-Driven Training

Let's list some of the most revolutionary advantages of using AI for team development:

1. **Quick Start** – No need to wait for the "perfect" training package. Start now.
2. **Always Available** – AI never sleeps. Create content at midnight or 5 a.m.
3. **Cost-Free** – No need to hire writers, editors, or video producers.
4. **Fully Customizable** – Change tone, structure, examples, and style.
5. **Scripture-Integrated** – Ask for verse-based teaching built right in.
6. **Faith-Aligned** – Guide the message with biblical principles at every step.

7. **Printable or Digital** – Share as Google Docs, PDFs, or email content.
8. **Modular** – Break long lessons into bite-size pieces that are easy to digest.
9. **Story-Driven** – Add parables, analogies, and real-life applications easily.
10. **Consistent Messaging** – Train dozens of people the same way, with the same quality.
11. **Personalized Pathways** – Ask AI to adjust for slow or fast learners.
12. **Instant Feedback** – Add quiz questions and reflection answers.
13. **Ongoing Coaching** – Create follow-up material without redoing anything.
14. **Scalable** – What you use for 3 people today can serve 3,000 tomorrow.
15. **Mentorship Ready** – Give mentors ready-made guides to use in 1-on-1s.
16. **Cross-Team Use** – Share training with other churches, teams, and networks.
17. **Time-Saving** – Spend more time building, less time teaching the same thing.
18. **Trackable** – Log what was taught, to whom, and when.
19. **Collaboration-Friendly** – Build it as a team, revise it together.
20. **Legacy-Driven** – Store it, repeat it, improve it, and pass it on.

This is not just practical—it's a picture of the Body working in unity with wisdom and speed.

Final Word: You Can Train Now—Without Delay, Without Limit

If you've ever said:

"I wish I had something to train people with…"
"I don't have time to teach this over and over…"
"We're growing too fast to keep up…"

Then this is your answer.

Use AI. Create the course. Format it in Google Docs. Name it clearly. Share it with the person who needs it. And keep building.

You don't have to wait until you're "ready." You're ready now.
You don't have to know how to build curriculum. AI can help.
You don't have to have perfect answers. You just need to care.

We're in a moment where **any willing builder can now equip others instantly.**

So why wait?

Start creating custom training for your Mutual Success Team today.
Save what you create. Share it. Refine it. And let your obedience become someone else's breakthrough.

This is how movements grow.

And it starts… one custom course at a time.

Chapter 30 – Learning That Sticks: Customizing Team Success Training for Each Unique Learner With AI

How to Teach and Learn Critical Skills—Even the Hard Stuff—Using Personalized AI Prompts That Make Everything Click

---

When Nothing Else Works—Try This

There are moments in every team where someone just isn't "getting it." They've been shown. They've been told. Maybe they've even tried. But the knowledge won't stick. The concepts won't click.

Maybe that person is you. Or maybe it's someone on your team who is faithful, willing, and committed—but stuck.

That's not a failure. It's a signal.
It means: *you need a different angle.*
And thankfully, with AI, you can get exactly that.

In this chapter, we explore one of the most **revolutionary and underused superpowers of AI**: **the ability to take a complex**

subject—and explain it in the metaphor of something you already love, already know, or already understand.

It's more than clever. It's deeply spiritual and wildly practical. Because when you learn something *in the language of your life*, it finally sticks.

So whether you're struggling to understand a new business strategy, a biblical principle, or a technical system—or trying to teach someone else something vital—**this one technique might be the key.**

The Learning Shift That Changes Everything

Let's say someone on your team needs to understand financial planning. But they've never taken a finance class in their life. You could explain profit margins and cash flow, and their eyes might glaze over.

But let's say you ask AI this:

"Explain how small business profit works, using a basketball team as the metaphor."

Now, your teammate who loves basketball suddenly understands:

- **Revenue = points scored**
- **Expenses = points allowed**
- **Profit = winning margin**
- **Budgeting = choosing your lineup based on opponent**
- **Cash reserves = your bench depth**

It lands. It sticks. It makes sense.
Because it's not random anymore—it's rooted in *what they already love.*

Or maybe you need to understand how prayer and consistency relate. So you ask:

"Explain a daily prayer life using the metaphor of growing a garden."

Suddenly, you see it. **Watering. Planting. Pruning. Time. Trust. Seasons.**
You feel it. You own it.

That's the power of customized learning through metaphor. And AI makes it possible, instantly.

Why This Is So Life-Changing for Mutual Success Teams

Every Mutual Success Team has people with different educational backgrounds, different learning styles, and different passions. Some are visual learners. Some are analytical. Some are emotional. Some are hands-on.

This technique levels the playing field. It means:

- No one gets left behind
- Everyone learns at their pace
- You remove shame from learning
- You accelerate skill development
- You turn complex systems into relatable ideas

It's not just about "teaching better." It's about building a culture where learning feels like *belonging*.

When someone feels like the teaching "speaks their language," they come alive.
They contribute.
They grow.

This is how we create *deep understanding* inside our teams—and raise up *competent leaders* who don't just follow... they own what they know.

How to Prompt AI for Customized Learning That Works

Here's a simple 3-step method to make this work for anything you need to teach—or learn:

Step 1: Name the Topic You're Trying to Learn
Example:

"Explain how to create and use a Google Doc to organize a church outreach."

Step 2: Add Your Personal Metaphor
Now tell AI how you want it explained. Use something you love.

"Explain this as if I'm learning how to run plays in a football game."
"Explain this like I'm preparing a recipe as a chef."
"Explain this like I'm building a Lego set with clear steps."
"Explain this as if it's a musical composition with different instruments."

Step 3: Ask for Depth and Clarity
You can then follow up with:

"Break it down into 3 clear phases, using that same metaphor."
"Now turn it into a 5-minute teaching with Bible examples."
"Summarize it for a teenager who learns visually."
"Add a step-by-step checklist using this example."

You've now created a **custom learning path** that actually fits how someone thinks—and ensures it *sticks for life.*

When the Lesson Is Too Important to Miss... Customize It

Sometimes you're learning something that's *too important to forget.*
It's critical. Strategic. Foundational.

But for whatever reason, it's not landing. It's not clicking.

That's your sign: **don't give up—customize the teaching.**

Use AI to:

- Help your kids understand tithing with a "video game" metaphor
- Teach your admin team project management using a "cooking show" model
- Explain Mutual Success principles through "music band dynamics"
- Break down website building using "gardening" language
- Relate team roles to "superhero teams" or "sports positions"

This isn't dumbing it down. This is *leveling it up.* Because now the learning sticks.

And when the learning sticks, people grow faster, teach others quicker, and become **owners** of what they once barely understood.

Final Word: Learning Should Be Joyful, Not Stressful

Jesus didn't teach with whiteboards and PowerPoints. He taught with *parables.* With stories. With metaphors that matched the audience.

Sowers. Sheep. Coins. Lamps. Builders. Farmers. Widows. Kings.

He taught through the life His listeners already understood.

And now you can do the same—thanks to AI.

You can help people on your team understand business, systems, leadership, software, faith, discipline, and outreach—all through *the filter of what they already love.*

So if someone's stuck—don't push harder. Don't make them feel dumb.
Just ask AI:

"Explain this as if they were a __________."

And watch the lights come on.

This is Kingdom education, in the 21st century.

And it starts right here—with one great metaphor, and one willing heart to learn.

Chapter 31 – The Always-Available Trainer: AI as an On-Demand Coach for Every Team Member

Why Every Person on Your Team Now Has 24/7 Access to Expert-Like Coaching—So You Never Have to Wait for Breakthrough Again

---

You Don't Have to Wait for an Expert Anymore

Have you ever needed help with something—something simple, something personal, or something urgent—and had no one to ask? Maybe the expert you know was busy. Maybe the person with answers was unavailable. Maybe you just didn't know who to go to.

Now, for the first time in history, that limitation is gone.

With AI, **you have access to an always-available, on-demand coach**—a source of guidance, brainstorming, encouragement, and support that can help you grow in just about any area of life.

No appointments needed. No schedules to manage. No waiting lists.
Just one question—and an answer comes back in seconds.

And while it's not a replacement for mentorship, discipleship, or the Holy Spirit—it is an **incredible supplemental coach** that is always there to help.
Any time. Anywhere. For anything.

The "Expert Coach" In Your Pocket—for 50+ Topics

Imagine this. You're trying to improve in one of the following areas:

- Time management
- Conflict resolution
- Focus
- Prayer routines
- Budgeting
- Outreach event planning
- Relationship boundaries
- Starting a new side business
- Studying the Bible
- Launching a product
- Writing a newsletter
- Fasting
- Creating accountability systems
- Organizing your files
- Building a new habit
- Delivering a speech
- Building a support team
- Overcoming procrastination

- Writing a vision statement
- Hosting a team meeting
- Learning how to ask better questions
- Helping someone through grief
- Making a decision with clarity
- Creating a job description
- Finding local Christian resources
- Building confidence
- Solving team problems
- Finding new donors
- Delegating more effectively
- Creating healthy boundaries

Guess what? **AI can help you with every single one.**

You can ask it:

"What are the top 3 ways I can improve in this area today?"
"Break this down into baby steps so I don't feel overwhelmed."
"Coach me like I'm a beginner and need encouragement along the way."
"Write this in the style of a spiritual mentor or wise business coach."
"Make it short, clear, and super practical."
"Encourage me at the end like I'm doing a great job."

The result? You feel equipped. Supported. Guided. And most importantly—**you're not stuck.**

How to Use the AI Coach to Strengthen Your Team Members Daily

Every person on your Mutual Success Team has areas where they're strong—and areas where they're struggling. And often, the reason they stall isn't lack of passion—it's **lack of personal coaching** in the moment when they need it most.

Here's how to change that using AI as your "always-on" coach.

Step 1: Identify a Real Challenge
Encourage your team members to ask, "What am I feeling stuck on today?"
Not in general, but *right now.*

Step 2: Turn It Into a Prompt
Train them to ask:

"Coach me through [topic] in 5 steps. Make it gentle, practical, and easy to start."

Examples:

"Coach me through overcoming fear of outreach."
"Coach me through organizing my day as a busy mom."
"Coach me through creating a vision for my small business."

Step 3: Take Action Immediately
What makes this tool powerful is not just reading the advice—it's *doing it.*
Most AI platforms will give immediate, step-by-step, custom-tailored advice. The user just has to try one small thing.

This is what makes AI transformational—not because it has "supernatural wisdom," but because it removes the **delay** that often keeps people from growing.

Why This Works So Well—Even If You're Not a "Coach" Type

Some people naturally lead. Others don't see themselves as coaches. But with AI, **everyone can be supported as if they had their own mentor.**

The best part? You don't have to "be the coach" all the time. Instead, you can teach your team to use AI to:

- Get clarity before team meetings
- Brainstorm outreach ideas
- Set goals for the week
- Solve interpersonal issues
- Write their own team updates
- Stay accountable to spiritual growth

This is how you multiply wisdom.
This is how you build maturity.
This is how you create a culture of self-growth and interdependence at the same time.

Everyone is getting coached. Everyone is growing. Everyone has access.

And you, as the team leader, are **no longer the bottleneck.**

Final Word: Coaching Is Now a Daily Gift—Not a Rare Event

There's nothing quite like hearing wisdom spoken directly to your situation, at the exact moment you need it.

That's what AI can now provide for you—and for every person on your team.

It's not about replacing human leadership. It's about *removing delays.*
It's about *keeping momentum.*
It's about *training people to ask better questions and take faster action.*

With AI as your on-demand coach, you can stop saying:

"I'll deal with this next week."
"I don't know where to start."
"I need to wait until someone helps me."

Instead, you can say:

"Let me get coached right now."
"Let me ask the right question and get unstuck."
"Let me build the life, the business, the team God called me to—today."

Because you are never without a voice of wisdom now.

AI can be your team's quiet, humble, 24/7 coach.
Always ready. Always kind. Always offering help.

And that, friends, is a tool worth using—every single day.

Part 7: People – The "Mutual Success Team" All Using AI

This part focuses on the people themselves—the real power behind every "Mutual Success Team." When every individual is equipped, trained, and able to use AI on their own, the team starts operating like a synchronized unit instead of a scattered collection of parts.

By encouraging everyone to engage with AI, the team begins to share more ideas, learn faster, and take ownership of the mission. AI becomes a unifying tool, helping people connect their personal faith with their team contributions. It creates more collaboration, more shared breakthroughs, and fewer delays.

Even those who feel like they don't know enough, or who have never led before, gain confidence when they see how accessible and helpful AI can be. It gives every person a meaningful place on the team—and every contribution can now be scaled.

When every voice matters and every member is equipped, you create unstoppable progress. This part teaches how to foster that environment, and makes clear that the success of any "Team Success Network" doesn't come from technology alone—it comes from Spirit-filled people using every tool God has made available.

Chapter 32 – Empowering the Whole Body of Christ: Letting Every Christian Contribute to the Growth of the "Team Success Network" With AI

How Each Believer Can Help End Global Need by Using AI and Hearing God—to Bless Others, Build Businesses, and Multiply the Kingdom

---

Yes—You Can Make a Real Contribution

You may not feel like an expert. You may not think you're a teacher, a builder, or a strategist. But make no mistake—**you have something to offer.**

God is raising up the Body of Christ not just as individual believers, but as **a unified network of Kingdom-minded doers**—those who care about each other's growth and are willing to invest their time, talents, and tools into something that serves more than just themselves.

That's the heart of the **Team Success Network.**

It's not about one person getting rich. It's about the **Mutual Success** of the whole Body.
It's about duplicatable businesses that lift up communities.
It's about digital tools—like AI and Google Docs—being used not just for efficiency, but for equipping.
It's about ending the phrase "there was need among them" in the Church, for good.

And the truth is: **AI can multiply any contribution you're willing to make.** Whether it's an idea, a document, a process, a testimony, or a piece of training—you can give something that becomes a blessing to someone else.

What You Have Can Help Someone Else

The vision of the Team Success Network is not just to create tools. It's to **empower believers everywhere**—in every city, country, and church—to build something sustainable, scalable, and successful.

But we can't do that alone. **We need your piece of the puzzle.**

Maybe you've figured out a better way to invite someone to a prayer meeting.
Maybe you created a flyer for your church's food drive.
Maybe you prompted AI to write a small business plan for your community, and it actually worked.

Save it. Share it. Send it in.
Let AI polish it. Let Google Docs organize it.
Let the Team Success Network **store it, distribute it, and bless others with it.**

Because here's the miracle: once you've made something once—with AI's help—it becomes easy to **duplicate, translate, adapt, and teach.**
That's multiplication.
That's Kingdom business.
That's God taking the fish and loaves in your hands—and feeding thousands.

AI + Inquiring of the Lord: Our Dual Superpower

We're not building this network based on technology alone. We are **Spirit-led**, always.

We're not just prompting AI—we're **inquiring of the Lord**.

When David faced major battles, he didn't just act. He **asked God first.** That's what we're building into every business decision, every outreach plan, every strategy.

We believe in:

- **Asking God before prompting AI**
- **Trusting the Holy Spirit to guide each step**
- **Checking our ideas against scripture**
- **Confirming things in prayer**
- **Walking forward in faith**

When you combine **God's voice with AI's speed**, you get something truly powerful.
It's not artificial wisdom—it's Spirit-guided innovation.

This is how we build companies that actually work.
This is how we support churches without burdening them.

This is how we make sure the solutions we create **carry Heaven's DNA**.

This Is About the Whole Body—And That Includes You

Let's be clear: God isn't just trying to bless a few Christians.
He wants the **entire Body** to be **fully supplied**.

In Heaven, there is:

- No lack
- No sickness
- No poverty
- No ignorance
- No division

Jesus said, *"The Kingdom of Heaven is near."*
And then He brought it near—through **teaching**, **preaching**, and **healing**.

Now it's our turn.

You have a smartphone? You can teach.
You have a testimony? You can preach.
You have a skill or business or idea? You can help someone be healed, freed, or lifted out of struggle.

And with AI, that contribution can be formatted, saved, shared, translated, and scaled.

This is what the "Team Success Network" is about.
Equipping you to be a part of something global, by doing something local.

Giving your wisdom, your wins, your systems—to bless others like you.

Final Word: Together, We Bring the Kingdom Near

We don't need another superstar.
We need the Body.
We need **you.**

God is asking us to support one another. To stop hoarding knowledge. To stop staying isolated. To stop believing that someone else will fix the problem.

He's giving us tools.
He's giving us each other.
He's giving us vision.
And yes—He's giving us AI.

So here's what you can do:

- Prompt AI to help you with something you've overcome
- Save the insight
- Format it as a short document
- Share it with the Team Success Network
- Keep doing it, as God leads

Each time you do, you're **helping bring an end to global need** in the Body of Christ.

You're helping us create a world where churches **lack nothing**, believers **thrive**, and the Kingdom is visible in **how we live and how we love**.

You're not too small. You're not too late.
You're part of this.
Let's bring Heaven near—together.

Want to send in your contribution? Email your documents or ideas to **docs@TeamSuccessNetwork.com**
Let's build this Kingdom library—together.

Chapter 33 – Helping to Support Discipleship – Personal & In Your "Mutual Success Team": How AI Can Play an Important Christian Role

Why AI Can Be a Practical, Indispensable Support Tool for Spirit-Led Discipleship—For Yourself, and for the Growth of Others in Your Team

---

Discipleship Needs Structure—and Support

Discipleship is the heartbeat of the Church. It's how faith grows from information to transformation. Jesus didn't just preach to crowds—He discipled a few, intentionally, personally, and consistently. And He told us to do the same.

But in our fast-paced, busy lives, real discipleship can be hard to maintain. We mean well—but people get overwhelmed. Leaders get exhausted. Mentors run out of bandwidth. And even those hungry to grow sometimes feel stuck.

That's why we must get creative. That's why we must think differently. And that's why **AI—used wisely—can become a**

powerful, Spirit-guided support for discipleship, both in your personal life and in your "Mutual Success Team."

It won't replace relationship. But it can support the structure. It can **provide consistency, clarity, and coaching** that helps everyone stay engaged in the process. Used in partnership with the Holy Spirit, AI can help create a discipleship rhythm that sticks—without adding burden.

Discipleship Isn't Optional—It's a Command

Jesus said, *"Go and make disciples of all nations."*
Not just converts. Not just followers. Disciples—*committed learners who are transformed over time.*

That's the mission of a healthy Mutual Success Team. It's not just about building a business or launching projects—it's about **helping each other grow in faith, character, purpose, and calling.**

And that takes:

- Ongoing teaching
- Practical tools
- Consistent encouragement
- Scripture engagement
- Personal reflection
- Goal tracking
- Conversations that go deeper

Now ask yourself: what if you had a free, always-available tool that could help with all of that?

You do. It's called AI.

AI can't disciple someone in spirit—but it can support **every natural part of the discipleship process.** It can help organize teachings, generate devotionals, suggest discussion questions, clarify biblical truths, and even help track growth steps over time.

Ways AI Can Help You Personally in Your Discipleship Journey

Let's start with you. Whether you're brand new to walking with God or have followed Him for decades, discipleship is still the path.

Here are a few ways AI can support your growth today:

1. Daily Word-Based Encouragement
Prompt: "Give me a short devotion and scripture to help me stay encouraged today in my walk with God."

2. Topical Bible Study Plans
Prompt: "Create a 7-day study on the fruits of the Spirit with reflection questions and application steps."

3. Personal Conviction Coaching
Prompt: "Help me reflect on how I can improve my time management as a disciple of Christ."

4. Scripture Memorization Help
Prompt: "Give me 5 memory tricks for memorizing Romans 8:28."

5. Thought Renewal Prompts
Prompt: "I struggle with comparison. What scriptures and affirmations can I use to renew my mind daily?"

6. Prayer Practice Coaching
Prompt: "Help me create a 10-minute daily prayer structure based on the Lord's Prayer."

7. Life Application Journal Templates
Prompt: "Give me a journal format to reflect on what God is teaching me each day."

8. Sermon Summary Tools
Prompt: "Help me organize and apply notes from today's church message."

9. Faith Goals and Action Steps
Prompt: "Help me break down my faith goals into weekly spiritual habits."

10. Tracking Growth Over Time
Create a shared document to log what you've been learning, praying, and applying—and let AI help summarize patterns.

When done consistently, this turns into **personal momentum**. It's not robotic. It's Spirit-led—with AI simply helping organize, prompt, and remind you of what matters most.

Ways AI Can Support Discipleship Across Your Mutual Success Team

Now imagine your whole team walking in that level of daily growth. It doesn't require more sermons—it requires more support.

AI can help make that happen.

1. Group Devotion Creation
Prompt: "Write a 5-minute devotional that ties our business to God's Kingdom."

2. Weekly Team Reflections
Prompt: "Give us 3 reflection questions for our team meeting based on Matthew 5."

3. Discipleship Action Plans
Prompt: "Create a 4-week growth plan for a new Christian joining our team."

4. Curriculum for Peer Mentorship
Prompt: "Help me write a one-on-one discipleship guide based on the book of James."

5. Conflict Coaching from Scripture
Prompt: "How can I handle a disagreement in my team biblically?"

6. Personalized Encouragement Texts
Prompt: "Write a short, kind, scripture-based message to encourage my teammate who's feeling discouraged."

7. Tracking Spiritual Progress
Create a shared doc with weekly check-ins on scripture reading, application, and prayer points.

8. Baptism and Faith Milestone Teaching
Prompt: "Give me a teaching plan on baptism for someone new in faith."

9. Vision Alignment Reminders
Prompt: "Write a simple paragraph tying our business progress this week to our Kingdom calling."

10. Multiplication Tools
Prompt: "Give me a simple discipleship template I can give to a new leader so they can disciple others."

With AI doing the **heavy lifting of structure**, you and your team are free to do the **heart lifting of relationship**.

Final Word: AI Can't Replace the Holy Spirit—But It Can Help Discipleship Stick

Let's be clear: nothing replaces **hearing God's voice**. Nothing replaces **the presence of the Holy Spirit**. And AI will never carry the anointing.

But that's not the point. The point is—**AI can clear the clutter**. It can organize your thoughts. It can suggest the next right question. It can make your lesson plan simple to follow. It can take the discipleship model you feel called to build—and give it a skeleton, a shape, a structure.

So you don't waste time. So people don't fall through the cracks. So your vision doesn't stay trapped in your head.

And the result?
People grow.
Faith deepens.
Obedience becomes easier.
Multiplication becomes natural.
The Great Commission becomes more than a mission statement—it becomes a way of life.

So use AI as your teaching assistant.
As your spiritual secretary.
As your outline-maker.
As your reflection prompt writer.
As your memory verse tutor.
As your strategic co-builder.

Because when paired with the Holy Spirit...
Discipleship becomes unstoppable.

Chapter 34 – Creating Content "NEEDED" on Command: Team Success Marketing, Flyers, Emails, and Social Posts

How to Use AI to Rapidly Generate the Right Words for the Right Purpose—And Get Better Results by Giving Better Direction

---

You Don't Have to Struggle With Content Anymore

Let's face it—most people don't love sitting down to write. Especially when it comes to creating **marketing material, outreach flyers, emails, or social media posts**. It feels like a chore. We second-guess ourselves. We don't know what to say or how to say it. And so, too often, we put it off—or worse, do nothing at all.

But in the world of Team Success and Mutual Success Teams, **communication is everything.**
If people don't know what you're building... they can't join you.
If people don't feel invited... they won't come.
If people don't hear your testimony, or your opportunity, or your vision... they stay disconnected.

This is where AI changes the game—because now, **you don't have to be a great writer.** You just have to be a clear communicator. If you can tell AI what you need, **you can get powerful, custom content in seconds.**

And not just "OK" content. You can get **flyers that feel anointed**, emails that stir faith, posts that create momentum, and language that **moves people to act**—toward your event, your offer, your ministry, your business, or your vision.

Here's How It Works: Prompt, Review, Refine

Let's say you want to create a flyer for a Saturday workshop your Mutual Success Team is hosting. You want it to sound exciting but spiritual. Clear but not too pushy.

Open AI (ChatGPT, Claude, Grok, Gemini) and type something like:

"Create a one-page flyer inviting Christian families to a Saturday morning workshop where they'll learn how to build a faith-based side business. Make it encouraging, clear, and include a verse."

Boom. AI responds with a draft.

Now you do three things:
1. Look it over.
Does it hit the right tone? Would YOU show up to this event based on this wording?

2. Ask for improvements.

"Make this sound more conversational."
"Can you include more emotional appeal and a stronger call to

action?"
"Use simpler language for someone without business experience."
"Add a subheading that makes people feel this is urgent."

3. Give better direction.
You'll notice—**the more specific you are, the better the result becomes.**
You can tell AI your target audience, the format, the goal, the emotion you want to stir, and even which scripture to include.

"Rewrite this for Instagram and include 3 short caption ideas."
"Make this email short, joyful, and sound like I'm talking to a friend."
"Summarize this into 3 points for a church bulletin announcement."

You're not stuck anymore. You're *equipped.*

What Kind of "NEEDED" Content Can You Create Instantly?

AI doesn't just help you write—it helps you *deliver what is needed most,* when it's needed most. And that's exactly how we live out the Kingdom value that "there was no need among them" (Acts 4:34).

Here's a list of just some of the content types you can create instantly with AI:

1. Outreach Flyers – Invite people to a service, seminar, healing event, or outreach
2. Fundraising Letters – Explain a need, share a vision, and call people to give

3. Social Posts – Promote a new launch, share testimonies, or invite engagement
4. Email Campaigns – Introduce your Mutual Success Team, business, or offer
5. Website Copy – Explain who you are, what you do, and why it matters
6. Presentation Slides – Teach a simple principle or pitch an idea
7. Blog Posts – Share your learning journey and inspire others
8. Devotional Emails – Encourage people with scripture + short reflection
9. Team Updates – Keep your group aligned, informed, and motivated
10. Product Descriptions – Make your offer clear, helpful, and spiritually grounded

Each one of these can be created with a simple prompt.
And with just a bit of refining, you'll sound like a pro—even if you're not.

How This Accelerates "No Need Among Them"

We're not just writing to sound good.
We're writing to **move people into action.**

Every time you clearly invite someone into:

- A church event
- A healing service
- A business opportunity
- A discipleship conversation
- A Mutual Success Team meeting
- A training or testimony

...you are helping meet a need.
And every time you remove confusion and replace it with clarity, you are lifting someone **closer to the provision God has for them.**

Here's the big vision:

"There was not a needy person among them."
That wasn't a metaphor. That was **a result of effective communication, shared resources, and mutual success.**

And now? With AI, you can do your part to bring that reality closer:

- By making clear offers
- By publishing persuasive words
- By creating invitations that feel warm and Spirit-led
- By sharing testimonies that build trust and spark faith
- By calling people into Kingdom work with the confidence of Heaven

---

Final Word: Speak Boldly—AI Will Help You Find the Words

Don't be afraid to write anymore.
Don't wait for the perfect moment.
Don't keep your business, your event, your vision hidden in vague language.

Instead, open up your AI tool, say what you need, **refine it together**, and publish with faith.

You're not just creating content.
You're **delivering Kingdom opportunity.**

You're helping someone find the answer they didn't know they were searching for.

And when you do this regularly—when your Mutual Success Team becomes known for **clarity, encouragement, action, and bold vision**—you're helping write the future of the Church.

One post, one flyer, one invite at a time.

So go ahead.
Create the content that's *needed.*
Let AI help you say it well.
And let the Kingdom grow, faster than ever.

Part 8: Relationships – Unity, Support – The Health of the Team

Strong relationships are the glue of every thriving "Mutual Success Team." This section teaches how AI can help strengthen those relationships—not by replacing personal connection, but by improving communication, empathy, and encouragement among members.

AI can help craft messages in a more loving tone, assist with hard conversations, and offer new ways to deliver grace and truth. It helps eliminate misunderstandings before they escalate. When team communication improves, unity follows—and that's when the Kingdom work gains traction.

In addition to communication, AI provides real-time encouragement. When someone needs a lift, a Scripture, a word of hope, or just a positive focus, AI is there to provide it. Spirit-led prompting helps you feed your team daily, not just weekly.

The overall health of the team is directly tied to the strength of its relationships. This part serves as a toolkit for emotional strength, communication clarity, and day-to-day peacekeeping. When members are seen, supported, and spiritually refreshed, the "Team Success" model moves from theory to a living example of God's grace.

Chapter 35 – Faster "Mutual" Team Building & Fundraising: Using AI to Help Your "Team Success"

How to Reach the Right People, Write the Right Words, and Grow the Right Support—With More Speed and Power Than Ever Before

---

Building Your Team Doesn't Have to Take Forever

One of the hardest parts of starting any Kingdom project is this: **finding the right people to build it with you.**
You might have the idea. You might have the heart. But without teammates, partners, and support—your vision stays stuck.

And right behind that challenge is another one: **funding the mission.**
How do you raise money for something meaningful, without begging? How do you communicate your heart in a way that moves others to give, to sow, to join?

For most Christians, these two areas—**team building and fundraising**—feel slow, awkward, and uncertain.

But now? There's good news.

AI can help you build and fund your "Mutual Success Team" **faster than ever before.**

We now have the tools to write the letters, craft the videos, clarify the message, and connect to the right people—**on command.**

And this chapter is your invitation to pull out all the stops. To expand your vision. To fully leverage AI in ways you may never have imagined—**to accelerate the building of your team and the funding of your mission.**

AI Can Help You Write Anything—For Anyone

Let's say you're starting your first Mutual Success Team. You've got the heart for it. You've got the vision. Now you just need:

- A short letter to invite a friend
- A welcome email for new members
- A fundraising message to raise $1,000
- A script for a 2-minute invitation video
- A presentation outline to share at church
- A caption for a social media post
- A landing page description for your project

This is where most people freeze.

But AI makes it simple.

Prompt:

"Write a short, encouraging invitation letter to a fellow Christian, asking them to pray about joining a new Mutual Success Team focused on Kingdom business. Include a vision statement and a clear call to action."

Refine:

"Make it more personal and warm, like it's coming from a friend who truly believes in them."

Convert:

"Turn this into a 2-minute video script with a natural tone and one scripture."

Multiply:

"Create a one-paragraph version for a text message and a longer version for an email."

You just created a complete outreach package—in minutes.

And now? You send it. You speak it. You share it.
And suddenly, **you're not building alone anymore.**

Fundraising Made Simple (And Spirit-Led)

Let's talk about money.

It's often awkward to ask for. But when your motive is pure—when your desire is to see God's people thrive and Kingdom projects succeed—**you need to get over the fear of fundraising.**

Because without resources, visions stall. And the Bible says, *"You have not, because you ask not."*

Here's how AI can help you raise support with confidence, clarity, and class.

Prompt:

"Write a letter to potential donors explaining the purpose of our Mutual Success Team, the impact we're making, and why we're raising $3,000 this month. Include a testimony and a way to give."

Then ask AI to:

- Shorten it into a powerful 5-sentence appeal
- Add a scripture about generosity
- Create a matching graphic caption for social media
- Turn it into a spoken-word-style video script
- Suggest a subject line for an email campaign

You can even ask:

"Give me 5 ideas for a fundraising video that feels joyful, not desperate."

"Help me write a giving page headline that communicates Kingdom ROI."

With every word, you're **removing confusion**, replacing hesitation, and making it easier for others to say "yes."

And as you do this... you're walking in obedience. You're helping end need.

Because fundraising done right isn't begging—it's **inviting people to sow into a future they want to be part of.**

Finding the Right People, Fast

Sometimes the hardest part of team building is knowing where to look.

AI can't magically find names and emails for you—but it can help you:

- Clarify your **ideal teammate profile**
- Write a list of **traits and values to look for**
- Create an **application or interest form**
- Draft a **referral message** for your network to send
- Plan a **team launch event** with AI helping you design the entire experience

Ask:

"Help me create a list of qualities I should look for in my first Mutual Success Team member."
"Write a short announcement I can share in a church group to find potential team members."
"Design a questionnaire that helps me assess someone's fit for our team culture."

Better yet, once people start showing interest, use AI to:

- Write welcome emails
- Draft follow-up messages
- Create quick-start training guides
- Generate encouraging check-ins

- Create team update templates

Team building becomes a process, not a mystery.

You're no longer waiting on someone else. You're equipped to move forward—today.

Final Word: The Body Builds Itself—But We Must Get Started

Ephesians 4:16 says the Body of Christ "builds itself up in love, as each part does its work."

That's exactly what this is about.

Your part is to act. To build. To reach out. To invite. To believe.

And now—with AI—you can:

- Communicate your vision with excellence
- Reach the people who were praying for a project like yours
- Invite support from those who want to sow into something meaningful
- Fundraise without fear or pressure
- Organize your outreach process
- Follow up with people quickly
- Create a launch plan that multiplies results

There are people waiting for you.
They just don't know about your team yet.
Let's change that.

Let AI be your scribe, your scriptwriter, your brainstorming partner, and your follow-up assistant.

You've got something worth building.
So don't let slow team formation or fundraising delay the mission.

Let's build together—faster, smarter, and Spirit-led.

Chapter 36 – Encouragement on Demand: Delivering Strength to Your Mutual Success Team Through Spirit-Led Prompts

How to Use AI to Speak Life, Faith, and Hope Into Your Team—Whenever They Need It Most

---

Sometimes All Someone Needs... Is a Word

We all hit moments when we're tired. When doubt creeps in. When the road ahead feels unclear. And in those moments, one timely word—one encouragement at just the right time—**can make all the difference.**

In fact, the Bible tells us this exact thing:

"A word fitly spoken is like apples of gold in settings of silver." (Proverbs 25:11)

That's what this chapter is about.

It's about creating **those fitly spoken words** for your Mutual Success Team.

Not once a year. Not once a month. But **on demand**—when they need it most.

Thanks to AI, you can now instantly generate Spirit-led encouragements for your team—crafted with scripture, filled with faith, and perfectly tuned to your team's current needs.

And when you start using this method, you'll find yourself stepping into a new level of leadership. You won't just be a coordinator. You'll be **a consistent wellspring of life-giving words**, helping your team thrive even when the road gets hard.

Encouragement Isn't Optional—It's a Kingdom Strategy

The enemy's goal is often very simple: **discourage you into stopping.**
Make you believe it's not working.
Make you feel alone.
Make you think your small obedience didn't matter.

But encouragement is a weapon. It pushes back darkness. It restores clarity. It stirs faith. It helps people keep going when everything says quit.

And here's the truth:

In your Mutual Success Team, **encouragement is one of your most valuable assets.**
If your people feel seen, valued, and uplifted—they'll stay. They'll grow. They'll multiply.

So don't wait until you "feel inspired."
With AI and the Holy Spirit together, you can start creating customized encouragement for your team—*on command.*

How to Generate Spirit-Led Encouragements With AI

You might be wondering: "How can a machine give a spiritual encouragement?"

The answer is: it can't—**by itself.**
But with *your faith*, *your prompting*, and *your direction*, AI can help you **format** the encouragement your team needs.

Let's break this down.

Step 1: Describe the Situation
Tell AI what's happening. Be specific.

"One of my team members feels discouraged because their first outreach didn't go as planned."
"Our team just launched a new product and we're not seeing results yet."
"I want to encourage everyone after a long week of building and serving."

Step 2: Ask for an Encouragement Prompt

"Write a 3-paragraph spiritual encouragement for this person, including one scripture, reminding them that progress takes time and God sees their effort."
"Give me a short text message to lift someone's faith and help them refocus on God's promises."

"Write a daily declaration for my team to say together tomorrow."

Step 3: Refine the Tone

"Make this warmer, more conversational, like it's coming from a friend."
"Add a touch of boldness—remind them of who they are in Christ."
"Include a quote from the Psalms."

Step 4: Deliver It Promptly
Don't wait. Send the message. Post it in your team chat. Read it aloud. Put it on your team flyer. Share it in person.

Encouragement delayed is encouragement denied.
With AI, you never have to delay again.

Creative Ways to Deliver Encouragement Using AI

There are endless ways to bring these Spirit-led prompts to life:

- **Morning Declarations**:
  Ask AI: "Create a declaration for Mutual Success Team members to speak over their day. Include scripture references."
  Deliver it via group text every morning.
- **Midweek Pick-Me-Up Posts**:
  Ask AI: "Write a short, faith-filled paragraph I can post midweek to lift my team's spirits."
  Add a custom hashtag and share.

- **Victory Celebration Messages**:
  When someone makes progress, prompt AI:

  > "Write a short celebration message for someone who made their first sale or reached a team milestone. Include a word of encouragement and a scripture."

- **Audio Messages or Voice Notes**:
  Take the message AI gave you, and **record it in your voice**. Add your heart. Speak life over your people.
- **Printable Encouragement Cards**:
  Ask AI to write encouragements you can print and bring to team meetings or mail out.
- **Personalized One-to-One Notes**:
  Ask AI:

  > "Write a short encouragement for someone who's balancing kids and work, and feels overwhelmed. Use Isaiah 40:31."

Whatever the method, **keep it Spirit-led. Keep it consistent. Keep it personal.**

Final Word: You Can Be the Voice That Lifts Someone Today

You don't need to be a pastor to speak into someone's life.
You just need to care.
You just need to be willing.
And now—with AI as your helper—you can be ready.

Encouragement isn't just a nice add-on.
It's a **Kingdom strategy**. It's a **team culture shaper**. It's a **seed of strength** that produces fruit in due season.

So go create:

- The next word that lifts a discouraged teammate
- The next reminder that God is still in control
- The next spark of faith that turns into bold action
- The next phrase that says "You're not alone—we're in this together"

Because the Body of Christ builds itself up in love.

And now, you've got the tools to deliver that love—**instantly, beautifully, and consistently.**

Encouragement isn't just possible on demand.
It's now available... through you.

Chapter 37 – The Benefits AI Can Offer When Communicating With Your "Mutual Success Team"

How to Strengthen Clarity, Consistency, and Care—by Letting AI Help You Communicate Like a Spirit-Led Pro

---

When Communication Improves—Everything Improves

No matter how strong your vision is… if your communication is weak, your team will struggle.
No matter how passionate your people are… if the messages aren't clear, consistent, and encouraging, things will fall through the cracks.

That's why communication is *the backbone of every successful "Mutual Success Team."*

Whether you're leading a project, organizing a prayer call, launching a new idea, or just encouraging the people walking beside you, **your ability to say the right thing in the right way—at the right time—can multiply your impact tenfold.**

And now, thanks to AI, you don't have to write every word on your own.
You don't have to guess what to say.
You don't have to wonder if your message will be understood.
You can **craft clear, encouraging, Spirit-aligned communication—on demand.**

In this chapter, you'll see exactly how AI can become your behind-the-scenes communication partner—helping you lead with peace, power, and precision.

---

10 Real Benefits of Using AI for Team Communication

Here's what changes when you start using AI for your internal team messages:

1. Clarity Without Delay
Instead of taking hours to write a group message or announcement, you can ask:

"Write a clear message to my Mutual Success Team explaining why we're shifting the meeting time and how it benefits everyone."

Boom. Clear, kind, confident communication—without the stress.

2. Encouragement That Uplifts
When someone needs encouragement, don't scramble. Ask AI:

"Write a 3-sentence encouragement to a team member who feels discouraged about slow progress. Include Romans 8:28."

3. Consistency Across the Team
AI can help you keep your language consistent across emails, texts, docs, and announcements—so your culture stays strong and aligned.

4. Tone Matching
Want your message to sound joyful? Gentle? Bold? Professional? AI lets you adjust the tone instantly.
Just say: "Make this sound like a spiritual big brother offering support."

5. Follow-Up and Reminders
Never forget to follow up again. Ask AI to generate polite, respectful reminders:

"Write a quick message checking in on whether they completed their part of the outreach flyer. Be kind and encouraging."

6. Meeting Summaries and Takeaways
Upload your meeting notes and prompt:

"Summarize this meeting into 5 bullet points with action items for the team."

7. Answering Questions with Wisdom
If a team member has a hard question, you can prompt AI to help:

"Help me respond to a teammate who's confused about why we use AI in our work. Make it biblical, clear, and affirming."

8. Preventing Misunderstandings
AI helps you word things in a way that feels gracious—not sharp or confusing—especially for tough conversations.

9. Saving Time on Repetitive Messages
Stop rewriting the same explanations. Use AI to draft evergreen messages you can tweak again and again.

10. Reaching Busy or Distant Members
If someone missed the meeting or hasn't been responsive, use AI to help craft a caring message that re-engages them with love and tact.

Every Part of Team Communication Can Be AI-Assisted

Here's just a glimpse of what you can prompt AI to help with inside your Mutual Success Team:

- Weekly team updates
- Meeting agendas
- Celebratory messages
- Decision announcements
- Vision recaps
- Conflict resolution messages
- Launch instructions
- Feedback requests
- Project progress reports
- Invitation letters
- Group devotional texts
- Message summaries
- "What's coming next" outlines
- "You're doing great" reminders

Even if you feel like communication isn't your strength—**it can be now.**

AI won't just make your messages faster. It'll make them *better*. More intentional. More caring. More effective.

Because now, you don't have to start from a blank page—you start with a clear prompt and **a Spirit-led purpose.**

How to Start Right Away: Simple Prompts That Work

Let's get practical. Try these plug-and-play prompts right now for your team:

"Write a quick Monday morning encouragement to energize our Mutual Success Team. Use one scripture, 3 sentences, and make it feel joyful."

"Help me write a message to thank someone who worked behind the scenes this week and made a huge difference."

"Summarize our outreach progress this week into a bullet-point update that's easy to read."

"Create a text reminder to bring food for Saturday's gathering, in a light and warm tone."

"Write a message to the team explaining our next step and asking for feedback."

"Give me a daily rhythm suggestion I can share with my team to help with time management and prayer."

"Craft a beautiful paragraph to restate our team's vision for ending need among Christians, and encourage unity."

You don't have to use them all at once.
But one great message a week—written with help—can **strengthen your team's culture more than 100 unspoken thoughts ever could.**

Final Word: Let AI Multiply Your Voice—Not Replace It

We're not using AI to do our job for us.
We're using it to **support our heart**, to shape our words, to simplify our message, and to **deliver love, clarity, and direction with excellence.**

Because your team matters.
Your words matter.
And communication isn't just a task—it's **a ministry.**

So whether you're sending a reminder or casting a bold vision...
Whether you're thanking one person or rallying ten...
Whether it's by email, text, social, or document...
AI can help you speak clearly, build culture, and reflect Christ.

This is how we honor the team God gave us.

So go ahead.
Send the message.
Deliver the encouragement.
Provide the clarity.

Let your words—strengthened by AI and saturated with the Spirit—be the fuel that helps your Mutual Success Team thrive, grow, and finish strong.

Chapter 38 – Strengthening Your Relationships: Practice Improved Communication in Team Success and Family Life

How to Use AI to Improve What You Say, How You Say It, and the Impact Your Words Have—In Both Ministry and Home

---

Better Communication Builds Stronger Relationships

Whether you're leading a "Mutual Success Team" or trying to grow a healthy family life, one truth holds: **your relationships rise or fall on how you communicate.**

It's not just about what you say.
It's how you say it.
It's the tone, the timing, the phrasing, the posture of the heart behind it all.

And most of us? We weren't taught this well.
We do our best—but sometimes our words miss the mark.
Sometimes we come off too strong.
Other times, not strong enough.

We wish we could go back and reword something. We wish we could say it better.

Now, thanks to AI, **you can practice and improve the way you speak, write, and even text—before the message ever goes out.**

With AI as your communication trainer, you can experiment with different tones, rewrite messages in more loving language, and ask for support crafting your words for **peace, power, and unity**—both in your team, and in your home.

AI as a Speaking Style Coach for Leaders and Loved Ones

Let's say you need to say something important—but sensitive. Something like:

- Giving someone helpful feedback
- Asking a teammate to follow through
- Bringing up a miscommunication
- Encouraging a spouse or child
- Writing a vision message to your team
- Correcting without offending

Instead of stressing or avoiding it, now you can go to AI and say:

"Help me say this in a warm, supportive tone: I need to remind a teammate they missed a deadline and it impacted the group."

Or:

"Write a message that thanks my spouse for how they've supported me this week, and make it sound deeply heartfelt."

Or even:

"Write a text I can send to my teenager that encourages them without sounding cheesy or preachy."

AI becomes a **real-time roleplay partner**, helping you rehearse and refine your tone.
You can ask:

- "Make this more gentle."
- "Add a note of gratitude."
- "Include a scripture to soften this."
- "Make this sound like a wise older brother."
- "Rewrite this like a kind mentor giving firm but loving advice."

This is how you **level up your communication maturity**—by being intentional before the moment hits.

The Tones and Styles You Can Learn to Use

Here are some powerful speaking styles you can ask AI to practice with you—and each one fits a unique situation in your life:

1. Supportive and Warm
Use this for relationship conversations where someone needs reassurance.

"You've been doing more than you realize, and I just want to say—I see it. I'm proud of you. You're not alone."

2. Firm but Loving
Use when correction is needed but your heart is for unity.

"I care about you and your role here. That's why I want to talk through what happened so we can get better together."

3. Passionate and Vision-Driven
Use for casting team vision and exciting others about the mission.

"We're building something the world hasn't seen—and you're a key part of it. Let's go further, together."

4. Peace-Filled and Calm
Use in conflict or tension, when emotions need soothing.

"Let's pause for a moment and remember: we're on the same team. Let's work this out in love."

5. Lighthearted and Joyful
Use when encouraging momentum or making people feel included.

"You brought the energy this week—thank you! Your presence makes a difference."

6. Biblical and Uplifting
Use when reminding others of God's promises.

"Remember, God is not done with you yet. Philippians 1:6 says He will finish what He started."

You can even ask AI to combine tones:

"Make this message sound passionate but respectful, with a touch of humor."

Over time, this becomes more than a tool—it becomes **a training ground for spiritual communication growth**.

Use AI to Practice Before Real Conversations

One of the best uses of AI is **rehearsal.**
You don't have to just *guess* how something will come across. You can practice.

Here's how:

1. **Describe the Situation**

"I need to talk to a teammate who's been missing meetings but I don't want them to feel ashamed."

2. **Ask AI to Draft Options**

"Give me 3 versions of how I could say this: one soft and empathetic, one direct but respectful, and one that includes a scripture."

3. **Review and Refine**
   Pick the tone that works best, then say:

"Can you make this a little more concise?"
"Add a word of encouragement at the end."
"Make it sound more like a pastor speaking in love."

4. **Practice Reading It Out Loud**
   The more you practice communicating with clarity and love, the more natural it becomes.

This isn't about sounding "robotic."
It's about being **thoughtful and Spirit-led** in how you steward your words.

Final Word: Communication Is Ministry—Let's Get Better at It

You may not think of yourself as a communicator.
But if you're building a team, raising a family, or leading in any way—**you are one.**

And that means how you speak matters.
How you write matters.
How you deliver a hard truth, share an invitation, or express your heart—it *all matters.*

So why not get support?

Why not let AI help you:

- Practice your tone
- Strengthen your writing
- Build your confidence
- Speak life into your family
- Build unity in your team
- Prevent misunderstandings
- Deliver feedback with grace
- Cast vision with fire
- Say "thank you" like it really means something

This is how we grow in love.
This is how we reflect Christ.
This is how we multiply harmony and mutual success across everything we touch.

So go ahead—let AI be your training partner in speech.
And let the Holy Spirit be your ultimate guide.

Together, they'll help you communicate like Jesus would—*full of grace and truth.*

Chapter 39 – AI Assisting With Difficult People and Situations: Staying Out of Trouble—Getting Expert Advice For Inside & Outside the Team

How to Handle Relational Challenges With Confidence and Wisdom—By Asking AI to Step Into Expert Roles When You Need Help Most

---

We All Face Difficult Situations—But Now, You Don't Have to Face Them Alone

If you're leading a Mutual Success Team, involved in ministry, or trying to build something Kingdom-minded in business—**people problems will eventually arise.**
Someone doesn't follow through.
Someone misinterprets your words.
Someone gets offended.
Or maybe… *you* don't know how to handle a sticky situation in the right tone or with enough grace.

The truth is, the way you **respond** can either make it worse—or resolve it in peace.

And in the past, you might have stayed up all night worrying... not knowing what to say, who to ask, or how to handle it.

But now? There's a new way to move forward wisely.

AI can help you work through difficult people and challenging conversations—by giving you expert-level feedback, guidance, and phrasing suggestions on demand.

This chapter is your guide to unlocking one of the most underestimated uses of AI: **getting real help with real issues—before they become real damage.**

AI Can Role-Play 7 Types of Experts to Help You Stay Out of Trouble

Sometimes we don't need someone to tell us what to do—we just need someone to help us **think better** about what we're about to do.
And that's exactly what AI can do—**especially when you assign it an "expert role."**

Here are 7 roles you can ask AI to take on:

1. Relationship Coach

"Act as a Christian relationship coach. I'm having trouble with a team member who feels overlooked. Help me phrase a message that rebuilds trust and opens the door to healing."

2. Communication Consultant

"You're now a tone specialist. Help me rewrite this email so it communicates clarity without sounding frustrated."

3. Legal Advisor (non-binding)

"You're a basic legal consultant. I had a contractor on our team break an agreement. What should I do next, and what precautions should I take?"

4. Conflict Resolution Expert

"You are a team dynamics coach. Give me 3 steps to de-escalate a situation between two team members who are arguing about leadership decisions."

5. Business Mentor

"You're a Christian business mentor. I have a client who hasn't paid. What's a godly, firm way to follow up without damaging the relationship?"

6. Spiritual Accountability Partner

"Act as a faith-based accountability coach. I'm feeling anger about someone's actions on my team. Help me process this through scripture and love."

7. Crisis Planner

"Help me prepare for a possible blowup. I'm worried someone might walk away from our project after a hard conversation. How can I prepare what to say and what to do next?"

By telling AI who it's playing as—you don't just get basic advice.
You get **insight tailored to the angle you need most.**
That's wisdom. That's safety. That's Kingdom intelligence.

How to Ask AI for Help With Your Problem

You don't have to phrase it perfectly.
You can literally start with:

"I have a problem. Help me with it."

But here's how to go deeper and get better help:

Step 1: Describe the Context

"I'm leading a 5-person Mutual Success Team. One of the members keeps going off-track in meetings and dominating the discussion. It's hurting the team unity."

Step 2: Ask for Help

"Can you help me decide how to bring this up in a loving, effective way?"

Step 3: Assign the Role

"Act like a pastor who leads group dynamics."
"Act like a professional business coach."
"Act like a therapist who values peace and directness."

Step 4: Request Multiple Options

"Give me 3 possible scripts I could use, with slightly different tones."
"Include a version that's more casual, and one that's more formal."

Step 5: Review & Pray

Ask the Holy Spirit to confirm what feels right.
Sometimes the best version is a blend between two drafts.

This process doesn't just help you resolve a situation.
It helps you **grow in how you handle people—before emotions explode.**

How This Applies to Family Life, Business, and Ministry

This isn't just for teams. These strategies can help in **any relationship**:

- **Family misunderstandings**

"Act as a marriage counselor. I want to explain to my spouse how I've been feeling overwhelmed without sounding like I'm blaming them."

- **Parenting conflict**

"You're a Christian parent coach. Help me write a message to my teenager about respecting our values."

- **Church tension**

"Act like a church elder. I want to ask my pastor for help, but I'm not sure how to say it respectfully."

- **Team disappointment**

"I need to apologize to my team for a delayed launch. Help me express accountability, and renew their hope."

- **Business breakdowns**

"I have to cancel a contract with a client. Write a message that is professional, godly, and gracious."

In each of these, AI doesn't replace the Holy Spirit.
But it helps you **test your words before they leave your mouth.**
That's wisdom—and it saves relationships.

Final Word: Don't React—Respond With Wisdom

Most problems get worse when we **react too fast, or say too little, or say it the wrong way.**
But with AI as your sounding board, and the Holy Spirit as your Guide—you can step into even the toughest moments with **calm, strategy, and grace.**

So don't try to wing it when emotions are high.
Don't carry the burden alone.
Don't feel like you're stuck without advice.

Instead, open up your AI tool, take a deep breath, and say:

"I'm facing something difficult. Help me figure this out with wisdom and love."

Then follow the prompts.
Pray as you review.
And move forward in peace.

Because now you're not just managing a moment.
You're leading through it—with maturity.
You're making peace possible.
You're preserving trust.
You're setting the tone for Team Success in every direction.

Part 9: Making An Impact

At the heart of everything is impact. This part highlights how AI tools can help believers not just organize and plan—but truly change lives. Whether it's personal healing, spiritual processing, or strategizing for church growth, the goal is transformation, not just productivity.

When someone is stuck emotionally, AI can help them journal, pray through pain, or discover scripture that comforts. When a church is trying to grow wisely, AI offers clarity, data, and planning support. When someone needs community resources, it helps find real-time, local help.

This part shows that AI is not cold and robotic—it's responsive, when prompted with care and wisdom. And when used with Holy Spirit discernment, it becomes a helper for life's most human moments: growth, grief, healing, and vision.

It leaves the reader encouraged: no matter how big the challenge or how small the team, help is available. Impact is possible. Real results can come from faith-filled planning, led by God and strengthened by tools we were meant to use. It's time to stop surviving and start building with power.

Chapter 40 – Healing, Processing, and Journaling: Using AI to Support Emotional Strength in Mutual Success Teams

How to Use AI as a Gentle Guide for Working Through Emotions, Building Resilience, and Staying Whole While Pursuing "Team Success"

Team Success Requires Emotional Strength

You may think a successful Mutual Success Team is built on ideas, strategy, and productivity. And yes—those matter. But if the people within the team are emotionally depleted, constantly overwhelmed, or carrying unprocessed pain, then **the foundation is fragile.**

Emotional strength is not just about being "tough." It's about being **resilient, honest, Spirit-led, and well-supported**—especially during challenging seasons of building.

The truth is, while you're building a team, launching a business, or organizing a Kingdom initiative, you're also:

- Processing disappointments
- Dealing with personal triggers
- Trying to stay encouraged
- Balancing family dynamics
- Navigating spiritual warfare
- Carrying emotional weight for others

You need a safe space to process. A place to think through your own emotions. Somewhere to explore your thoughts—without judgment, without pressure, and without needing someone else to always be available.

And now, thanks to AI, you have that.
AI can serve as a private journal assistant, a compassionate reflection tool, and a gentle coach to help you process and heal in real time.

Why Emotional Processing Helps Your Mission Thrive

Let's say you have a hard team conversation, and you leave feeling discouraged. Or maybe your project hits a delay. Or someone misjudges your motives. Or a family situation pulls your focus—and you start questioning your ability to lead at all.

It's easy to suppress those feelings.

But what happens when you do?
You show up with less clarity. You speak with tension in your voice. You struggle to focus. You delay action—not because you're lazy, but because you're carrying invisible weight.

That's why Jesus often withdrew to pray. He wasn't avoiding people. He was **processing with the Father**—emotionally and spiritually. And we're invited to do the same.

AI can support this process in a surprisingly practical way.

When you say:

"I'm feeling stuck. Help me unpack this."

Or:

"I had a discouraging call with a teammate. Help me reflect on it with grace."

AI can walk you through:

- Gentle questions
- Scriptural encouragement
- Journaling prompts
- Clarifying summaries
- Emotional validation
- Mental reframing

It becomes a mirror—not to replace your faith or relationships—but to help you **slow down and hear yourself again.**

Prompts That Support Healing and Clarity

Here are examples of real prompts you can give AI to walk you through emotional strengthening and inner healing:

1. Journaling Your Thoughts

"Act as a journaling guide. Help me reflect on what I'm feeling today. Ask 5 questions that will help me understand what's under the surface."

2. Processing a Tough Day

"Help me work through frustration I felt today after a team member ignored my input. I want to process this with wisdom, not bitterness."

3. Dealing With Internal Pressure

"I feel behind and overwhelmed. Help me name these emotions, and remind me what God says about carrying burdens."

4. Releasing Unforgiveness

"Lead me through a reflection to forgive someone who let me down. I want to obey God, but I feel stuck."

5. Preparing for a Difficult Conversation

"I need help calming my thoughts before confronting someone. Walk me through a mindset of peace and humility."

6. Grieving a Loss

"Help me process a personal loss I haven't fully dealt with yet. Gently ask me reflection questions, and guide me toward healing."

7. Rebuilding Confidence

"I feel like I failed this week. Help me reframe my thoughts and speak truth over myself."

In each of these, AI becomes a tool—not to "fix" you—but to **slow you down, let the Spirit speak, and bring your soul into peace again.**

Why This Matters for the Whole Team

When *you* are emotionally healthy, your **entire Mutual Success Team benefits.**

You'll:

- Speak more clearly
- React with more patience
- Handle pressure with grace
- Recognize when others are struggling
- Be less defensive in conflict
- Stay steady in storms
- Discern decisions with clarity
- Lead with a soft heart and a strong mind

Now imagine **every team member** using AI like this:

- Journaling quietly before a team meeting
- Reflecting after a stressful day
- Writing a gratitude log
- Releasing offense before it poisons momentum
- Getting coached through imposter syndrome
- Rewriting a negative narrative into a faith-based one

You create a culture of self-awareness, healing, and **personal responsibility.**
And that becomes a team marked by **emotional integrity.**

This is what healthy Kingdom leadership looks like—not just passion and productivity, but **peace and maturity**.

Final Word: Let God Use Every Tool to Strengthen You

At the end of the day, AI isn't a counselor, a pastor, or the Holy Spirit.

But it *can* be a gentle mirror.
It *can* ask better questions.
It *can* give you space to process when others aren't available.
It *can* help you slow your mind and listen to your heart.

And most importantly—it helps you **hear God more clearly**, because you've made space to listen.

So next time you feel heavy, foggy, overwhelmed, or off-center, don't just push through.

Pause.
Open your AI tool.
And say:

"Help me reflect on this with peace. Help me hear God's voice again."

That single practice may be the key to unlocking your next level of clarity, courage, and strength—for your sake, and for your team's.

Because healing isn't just about relief.
It's about becoming **ready**—for the next breakthrough, the next mission, the next miracle.

So go ahead—process, journal, reflect, and receive.
You'll rise stronger, wiser, and clearer than ever before.

Chapter 41 – AI for Strategic Church Growth: Letting Technology Support the Expansion of the Body of Christ

How Artificial Intelligence Can Help Churches Grow Stronger, Reach Further, and Equip the Body More Effectively—Without Losing the Spirit

---

The Church Was Always Meant to Grow

From the moment Jesus said, *"Go and make disciples of all nations,"* the Church was called to grow—**not just in numbers, but in impact.** The early Church was marked by teaching, outreach, miracles, fellowship, and multiplication. And today, the mission hasn't changed. But the tools have.

We're living in an era of **unprecedented access to technology**, and that includes the rise of Artificial Intelligence. For many pastors, church leaders, and ministry workers, AI might feel intimidating—or even irrelevant. But the truth is, **AI is not a replacement for the Holy Spirit—it's a support system to help the Body of Christ expand.**

This chapter is about **using AI wisely, prayerfully, and strategically** to help your church or ministry grow stronger, reach more people, train better leaders, and multiply the work of God—faster and with less burnout.

You don't need to become a tech expert. You just need to become open to partnering with the tools God has made available—**tools that, when guided by faith and discernment, can radically increase your Kingdom impact.**

What Strategic Growth Actually Looks Like

When we talk about "strategic church growth," we're not just talking about filling seats. We're talking about:

- Equipping volunteers faster
- Following up with new visitors effectively
- Communicating with clarity and consistency
- Offering training and courses that are accessible 24/7
- Documenting the move of God so it can be multiplied
- Creating resources that strengthen the whole Body

And these are all areas where **AI can help.** Not by replacing humans. But by empowering them to do more—with less strain.

Here are just a few areas of church operations where AI can bring massive value:

1. Visitor Follow-Up

Prompt: "Write a warm, Spirit-led follow-up message for someone who visited our church for the first time. Include a scripture and invitation to a small group."

2. Volunteer Training

Prompt: "Create a 5-part training plan for new greeters at a Sunday service. Keep it joyful and aligned with our church culture."

3. Pastor's Weekly Email

Prompt: "Summarize Sunday's sermon into a weekly email for members. Include one reflection question and one verse to meditate on."

4. Prayer List Organization

Prompt: "Organize this prayer list into categories (healing, financial need, family) and suggest encouraging scriptures for each."

5. Sermon Planning

Prompt: "Help outline a 4-week sermon series on 'Walking in God's Purpose.' Include title suggestions and core verses."

All of these are tasks a human still oversees—but **AI accelerates the prep work** so more time can be spent in ministry, prayer, and people care.

AI Helps Your Church Become More Reproducible

One of the greatest challenges in ministry is this:

How do we make what God is doing here available to others somewhere else?

We've seen churches experience revival... but not document the steps.
We've seen ministries birth life-changing programs... but not train anyone else how to start them.
We've seen outreach strategies work in one city... but remain unknown in the next.

AI allows your church to:

- **Document** your strategies
- **Summarize** your playbooks
- **Organize** your wisdom
- **Translate** your materials
- **Convert** spoken stories into teaching outlines
- **Repurpose** sermons into devotionals, blogs, and books

You can even ask:

"Turn this transcript from our healing service into a 10-point volunteer guide on how to pray for others."

This turns every win into a training.
Every message into a seed.
Every breakthrough into a blueprint.

And that's how **you don't just grow one church—you multiply Kingdom effectiveness across the Body.**

Creating a Culture of Innovation—Not Fear

Many leaders are hesitant about AI because they worry:

- It might be misused.
- It might make things too easy.
- It might replace personal interaction.
- It might feel "unspiritual."

Let's be clear: **AI is a tool.** It's not your pastor. It's not your prayer team. It's not your Holy Spirit.

But when used wisely and prayerfully, it becomes a **faithful assistant** to help your team:

- Work more efficiently
- Plan more strategically
- Communicate more clearly
- Train more effectively
- Organize more intentionally

It becomes the "scribe" that records your vision. The "project manager" that reminds you of follow-ups. The "assistant" who never sleeps, always types, and makes your job a little lighter—**so your hands are freed up to lay hands, your time is freed up to pray, and your heart is freed up to love well.**

Innovation doesn't mean compromise.
It means moving forward in **partnership with the Spirit—and the tools He's made available in this time.**

Final Word: Expand the Body, Empower the People, Elevate the Vision

God's heart is for growth. For multiplication. For more laborers in the field. For churches that are thriving—not just surviving.

AI won't grow your church for you.
But it can help you grow with less stress, more organization, and greater reach.

Here's how to start:

- Choose one area where communication or planning feels heavy.
- Ask AI to help lighten the load.
- Pray as you read the output.
- Use discernment and revise what needs refining.
- Test it. Share it. Improve it. Repeat it.

Before long, you'll start seeing the fruit:

- More engaged members
- More equipped volunteers
- More consistent communication
- More souls reached
- More testimonies documented
- More of your time going where it truly matters

This is how we **expand the Body of Christ—without burning out the Body.**

You are called to grow.
So let's grow *strategically.*
Let's grow *together.*
And let's use every tool God allows to strengthen the Church He's coming back for.

Chapter 42 – Finding Local Help: Using AI to Identify Community Resources and Faith-Based Services

How to Use AI's Search Features to Uncover Local Solutions, Churches, Ministries, and Support—For You and Your "Mutual Success Team"

There's Help Nearby—If You Know How to Look

One of the greatest frustrations Christians face is this:
You need help—real help—but you don't know who to call, where to go, or what's even available in your local area.

You might be looking for:

- A ministry that delivers groceries to families in need
- A job skills program that helps people get hired
- A faith-based addiction recovery group
- Churches offering mental health or prayer counseling
- A Christian business network for support and training
- Local organizations offering free legal, housing, or medical help

Often, these resources **already exist**—but they're scattered, poorly advertised, or buried in outdated websites. Many churches and ministries aren't visible through a typical search engine.

That's where AI becomes a powerful ally.

Using the search features within tools like ChatGPT, Claude, Gemini, or Grok, you can **ask the right questions in the right way**—and uncover real local help in minutes.

This chapter will walk you through **how to prompt AI to find community-based and faith-based resources**, so you never have to walk alone—or let someone in your Mutual Success Team suffer in silence.

What Makes AI Search Different—And Better

When most people think of searching online, they think of Google. And while Google is useful, it's not conversational. It gives you links—not guidance. You still have to sort through page after page and guess which one is relevant.

But when you use **AI with live search capabilities**, you're not just typing keywords. You're **starting a dialogue.** You're saying:

"Hey AI, help me find a free Christian counseling center within 15 miles of me."
"Help me locate ministries in my city that help with job training."
"I need a food pantry that's open on weekends and is connected to a church."
"Find Christian addiction recovery groups in my zip code that meet weekly."

"Are there local Christian business mentorship groups or entrepreneurs' networks?"

When you phrase your question like this, AI doesn't just point you to generic sites—it actually filters and refines information to match **your need, your location, and your intent.**

And even if one AI tool doesn't produce a helpful result, another one might.

Use All Four AIs: Multiply Your Search Power

To truly unlock the potential of local help, **don't rely on just one AI.**
Each AI platform has its own way of searching, pulling sources, and interpreting questions.

Here's a quick breakdown:

1. ChatGPT (with Browsing)

Best at: Concise answers with source links, gentle tone, good summarization
Prompt:

"Search for churches in Atlanta that offer weekday healing services or inner healing prayer ministries."

2. Claude AI

Best at: Long-form reasoning, detailed summaries, sorting through dense data
Prompt:

"Compare three Christian ministries in my area that provide transitional housing support. Include pros and cons."

3. Google Gemini

Best at: Real-time Google data + map-based results
Prompt:

"List top-rated Christian family counseling centers near Phoenix, AZ. Include contact info and website links."

4. xAI Grok (Twitter-integrated)

Best at: Trend detection, crowd-sourced info, newer initiatives
Prompt:

"Are there any new Christian entrepreneur events happening in my city this month?"

Use them **all together.**
Same prompt. Copy and paste. Run it across each tool.
Compare the results.

This one simple habit can uncover resources, solutions, and ministries that would've taken you days—or never appeared at all.

Local Help You Didn't Know Existed—Until Now

Here's a sample list of categories where AI can help you uncover local help:

- Food banks & meal programs (church-based)
- Homeless shelters with spiritual support
- Employment services run by Christian nonprofits
- Parenting support groups in local churches
- Volunteer opportunities in faith-based orgs
- Free or low-cost Christian counseling
- Disaster relief ministries operating nearby
- Churches with business networking breakfasts
- Christian schools with tuition assistance
- Ministries offering utility or rent help
- Grief support groups hosted by churches
- Christian legal aid or mediation services
- Bilingual ministries for immigrant communities
- Pro-life pregnancy centers offering free support
- Prayer walk events, worship nights, or healing crusades

The list goes on.
But the power lies in how you **search.**

Be specific. Use city names. Include keywords like "Christian," "faith-based," "ministry," or "church."
And don't forget to follow up with:

"Can you give me the address, phone number, website, and service times?"

Final Word: Your Team Is Stronger When Needs Are Met

Your Mutual Success Team is built on shared progress. But sometimes, people get stuck because of **practical, solvable needs**—and they're too afraid to ask, or they assume no help exists.

This is your opportunity to change that.

You now have a way to quietly, kindly, and persistently seek out:

- Housing help
- Food solutions
- Healing opportunities
- Skill-building ministries
- Spiritual restoration spaces
- Job placement programs
- Encouraging, faith-filled support groups

You can do it **without judgment**.
You can do it **on your phone**.
You can do it **any day of the week.**

So whether it's for you, your team, your church, or a stranger you just met—make AI a part of your **Kingdom search strategy.**

Because someone's miracle might already be happening... just a few streets away.
And now, you'll know how to find it.

Part 10: Our Projects – "Mutual Success Team" Help

The final section is where everything comes together. It's not theory anymore—it's building time. These chapters guide readers in using AI to support actual "Mutual Success Projects" from start to scale. It's the blueprint for real Kingdom work, made repeatable, trackable, and shareable.

Workflows are created. Documents are duplicated. Operations are sped up. This part emphasizes how AI can help standardize success so that it can be duplicated across churches, teams, and nations. No need to start from scratch. Start with what works—and let it scale.

Readers are encouraged to document breakthroughs, share their working methods, and contribute to the wider "Team Success Network." By sharing what works, everyone benefits. And as more teams submit their resources, we move toward the greater mission: that there would be "no need among them."

By the time readers finish this section, they realize they've been empowered not just to learn—but to lead. The final chapters are a launchpad for long-term impact. They serve as a call to action, a final nudge to go out and build boldly, with faith and the tools God has placed in our hands.

Chapter 43 – Creating Repeatable Success: AI-Powered Documentation for Duplicating Mutual Success Projects

How to Use AI to Record, Refine, and Replicate the Best of What Works—So Your Team's Success Can Multiply Across the Body of Christ

---

You've Done Something That Worked—Now Make It Duplicatable

One of the greatest missed opportunities in ministry, business, and even Mutual Success Teams is this:

We succeed... and then forget how we did it.

A team launches a fundraiser that works.
A family organizes a support system that relieves stress.
A local church runs a community outreach that brings in new people.

But once it's over, nobody wrote it down. Nobody captured the steps. Nobody documented the journey.

And the result?
It becomes a one-time miracle instead of a repeatable model.

But that's not how God builds movements.
He builds through *patterns*—through processes that can be passed down, shared, and spread.

That's why in this chapter, we'll show you how to use AI to create **repeatable documentation** of your Mutual Success Projects—so that what works for one team... can work for many.

Why Documentation Is One of the Most Powerful Things You Can Do

Let's imagine a breakthrough in your team:
You find a strategy that generates consistent weekly income... or builds a strong prayer team... or helps struggling families stabilize. Praise God!

Now imagine if:

- Another church could follow your exact process and get the same result.
- A ministry leader across the country could read your plan and launch it in their own city.
- A Christian business leader could take your 12-step playbook and build something similar with their own group.

This is what happens when we move from **random effort** to **documented process.**

And AI makes this incredibly easy.

You don't need to be a writer. You don't need to have a background in project management.
You just need to ask AI to **help you turn what you've done into something that others can follow.**

This is how your impact multiplies.
This is how you create Kingdom blueprints.
This is how you build the Body of Christ beyond your borders.

How to Use AI to Create Project Documentation

Start with this simple prompt:

"Help me create a step-by-step guide for a Mutual Success Team project I just completed. We ran a 3-week outreach campaign in our city, and here's what we did..."

Then list your rough notes—whatever you can remember:

- Dates
- Team roles
- Tasks
- Outcomes
- Resources used
- Challenges faced
- Key decisions made
- Scriptures that guided you
- Testimonies from people involved
- What you'd do differently next time

Now ask AI to:

"Turn this into a clean, readable project report I can share with other churches or teams. Include steps, tips, and spiritual insights."

In minutes, AI can generate:

- A project summary
- A timeline
- A checklist
- A list of tools used
- Reflections and suggestions
- Scriptural foundation
- Downloadable guides and templates

Now your project isn't just a memory.
It's **a movement starter.**

Examples of Repeatable Success Projects You Can Document

Think about what your Mutual Success Team has done—or is working on right now. Here are some common categories worth documenting:

- **Fundraisers**

  "How we raised $1,200 in 14 days to cover a team member's car repair."

- **Outreach Events**

  "How we planned a church block party that brought 100 new visitors."

- **Team Routines**

"How we use Monday AI meetings to plan our week in 20 minutes."

- **Bible Courses**

"Our 5-day scripture study flow using AI chapter prompts."

- **Social Media**

"How we use AI to schedule 10 days of content in 30 minutes."

- **Resource Sharing**

"How we organized 5 churches to share food pantry inventory."

- **Training Sessions**

"How we trained new team members using AI-generated flashcards."

- **Healing Prayer Journals**

"How we documented 10 healing testimonies using AI and shared them globally."

Even if you think what you've done is small—**it's worth writing down.**
Because what seems simple to you might be exactly what someone else is praying for.

Final Word: God Uses Patterns—Let's Capture Ours

The Bible is filled with patterns and processes.
God gave Noah exact measurements.
He gave Moses the blueprint for the tabernacle.
Jesus gave the disciples a commission.
Paul gave Timothy instructions for building churches.

Why?
Because **what works in the Spirit needs to be passed on.**

AI is your tool for this moment.
To help you write it down.
To help you remember what God showed you.
To help you clarify how it came together.
To help others replicate what you've done—*with their own anointing.*

You don't need perfection.
You need to capture the process.

So next time something good happens, don't just celebrate it.
Prompt AI and say:

"Help me turn this success into a repeatable guide."

And then?
Share it.
Post it.
Email it to another team.
Send it to the Team Success Network.

Because together, we're building a library of solutions.
A Kingdom archive of faith-in-action.
A practical roadmap toward **"no need among them."**

And it starts with what you've already done.
Let's multiply it—one documented project at a time.

Chapter 45 – The Digital Assistant That Never Sleeps: Using AI in Your "Team Success" Operations

An Expert Who Is Never Annoyed, Never Tired, and Always Ready to Help You Automate, Speed Up, and Duplicate Mutual Success Projects

---

You Don't Have to Do It All Alone Anymore

In every Mutual Success Team, there are dreams. There are goals. There are people with passion. But there's also a reality—**time is limited, energy fluctuates, and people get overwhelmed.** Some days it feels like no one is available. Your team might be tired, busy, unsure what to do next, or struggling to keep up.

But what if you had a helper—someone who:

- Never forgets
- Never gets tired
- Never rolls their eyes
- Works 24/7
- Responds instantly
- Keeps everything organized

- Offers new ideas
- And never stops showing up

That's what AI is. **A digital assistant for your Mutual Success Team who never sleeps and is always ready.**

This chapter will show you how to use AI as your behind-the-scenes team member, helping you **automate tasks, speed up operations, and create duplicatable projects**—so that what you build can grow faster, reach more people, and require less stress to maintain.

AI Can Now Do What Used to Take 10 Volunteers

Let's say you're trying to:

- Plan your next team meeting
- Write the email invite
- Create a quick flyer
- Send out social media posts
- Write a prayer follow-up
- Summarize a past training
- Build a task checklist
- Organize your project doc
- Design a simple survey
- Create a follow-up letter

Normally, that's a week of work—or 5 people's job.
But now?

You can open ChatGPT, Claude, Gemini, or Grok and say:

"Create a flyer for our Mutual Success Team's upcoming strategy night. Include the date, theme, and scripture: Habakkuk 2:2."

"Write an encouraging email to the team with 3 bullet points from our last session and one action item for this week."

"Make a checklist for onboarding a new member into our Christian Decks outreach project."

"Write a follow-up message for someone who received prayer last week, reminding them we're still standing with them."

Each time you do this, you're tapping into an **always-on digital assistant** who takes your needs, runs with your instructions, and delivers results in seconds.

And the best part? It gets better the more you use it.

Here's How to Make It Work for Your Team

To truly benefit from your "AI assistant," follow this workflow:

1. Name the Project

Get specific. Is this outreach? A fundraiser? Training?

"We're planning a 3-week invite campaign for new families."

2. Assign the Assistant Tasks

Tell AI what you'd normally ask a volunteer.

"Create a social post for each of the 3 weeks with a scripture, a call to action, and a kind tone."

"Generate a list of FAQs people might have about the event and short answers we can post."

"Write a one-paragraph announcement for our church bulletin."

"Create a follow-up message template to send 3 days after the event."

3. Organize and Save

Once you've generated these pieces, **save them in a shared Google Doc** labeled by project.

Now, you have an archive of:

- Reusable templates
- Past messages
- Full content packages
- Success plans ready to duplicate or improve

And you can hand this doc to anyone new and say,

"This is how we did it. Just run the same play."

This is how **Mutual Success becomes reproducible.**
This is how churches in other cities can learn from what worked for you.
This is how one good thing becomes many great things.

Why This Makes Your Team So Much More Effective

With your AI assistant in place, you now have:

- **Speed:** What took hours takes minutes.
- **Consistency:** Your tone and brand stay strong across all platforms.
- **Confidence:** No more wondering "how do I say this?"—AI drafts it first.
- **Organization:** Documents and plans are no longer in your head—they're saved and ready to share.
- **Relief:** You don't have to do it all. You just have to delegate to AI and refine what it gives.

This doesn't just help you execute better—it **frees your team up to focus on people.**
More time for prayer. More time for one-on-one care. More time for Spirit-led decisions.

Because your assistant is handling the background.

Final Word: What Could You Build—If You Had Help Every Day?

Every leader dreams of having a team that's "on it."
One that responds quickly. One that shows up early. One that brings new ideas.
Now imagine if **you had that level of support, 24 hours a day—just one click away.**

That's what AI offers. Not magic. Not mind-reading. But **faithful, instant assistance** that can help you move faster, communicate better, and structure your projects with excellence.

Your Mutual Success Team doesn't have to wait until it's big to be effective.
You can start small—and move fast—**when you use AI as your silent team member.**

So start asking for help.
Assign tasks.
Give directions.
Test results.
Save your best outputs.
Build your operations library.

And the next time someone says,

"How are you guys doing so much with so few people?"

You'll smile and say:

"We have a helper. An expert who never sleeps—and always delivers."

And quietly, you'll know:
You've learned how to multiply with wisdom, scale with simplicity, and build Kingdom impact... one AI-generated step at a time.

Chapter 46 – Documents That Duplicate: Letting AI Help Your "Team Success" Projects Multiply Across Churches

Using AI to Create Easy-to-Share Templates and Process Guides So Other Teams Can Succeed Just Like Yours

---

What If Your Team's Breakthrough Could Become Someone Else's Blueprint?

Most of the time, when a "Mutual Success Team" creates a project that works—whether it's an outreach, a business launch, or a church-led ministry—something amazing happens.

But only once.

Why?
Because no one wrote it down.
No one turned it into a repeatable guide.
No one created the kind of document that another church, ministry, or team could take and say:

"Let's run this same thing here and watch God move."

This chapter is about fixing that.
It's about making your wins **transferable.**
It's about using AI to build the kind of documents that help **other people duplicate your success.**

When you do this, you move from isolated breakthroughs... to **networked multiplication.**
And that's when things get powerful.

Duplication Is a Kingdom Principle

In Acts, when churches were planted, they didn't start from scratch every time. They followed a pattern. A framework. A reproducible structure.

Paul even wrote detailed letters to clarify the processes:

- How to appoint leaders
- How to handle conflict
- How to manage offerings
- How to teach new believers
- How to stay aligned with the Gospel

Why? Because he understood something: **Duplication leads to multiplication.**

Today, we're being called to do the same. Not just to build something amazing—but to **document it, structure it, and share it**—so it can happen again and again.

And now, with AI, this has never been easier.

You can take your successful Mutual Success Team project and turn it into:

- A step-by-step startup guide
- A downloadable checklist
- A team training overview
- A communication template
- A budget sample
- A schedule layout
- A marketing plan
- A scripture-based mission doc
- A full vision presentation deck

All created quickly... all ready to be duplicated.

How to Turn a One-Time Project Into a Duplicatable Document

Here's a real-world example.

Let's say your team launched a Christian Decks scripture card business and got it to $1,000/month in revenue.

Instead of just celebrating it and moving on—ask AI:

"Turn our project notes into a step-by-step startup plan other Mutual Success Teams can follow."

Then feed it:

- Your startup story
- What steps you followed
- What tools you used
- What decisions you had to make

- What obstacles you faced
- What your timeline was
- What scriptures or prayers guided you
- What lessons you learned
- What you'd do differently next time

AI can then create:

- A formatted guide (with bullet points, sections, and tips)
- A downloadable Google Doc or PDF
- A training script or teaching outline
- A checklist for weekly operations
- Templates for emails, texts, or flyers
- Suggested job roles and time expectations
- Visual summaries and workflows

Now your one-time win is **a repeatable resource.**
You can share it freely.
Post it. Teach it. Send it to the Team Success Network.
And watch as others build off your foundation.

How This Helps the Whole Body of Christ Multiply Faster

Imagine this future:

- A Mutual Success Team in Ohio hears about your Christian Decks success.
- They download your document.
- They follow the steps.
- They reach the same income milestone.
- They create a new flyer using your AI template.
- They modify the launch prayer you shared.

- They host a local training using your outline.

Now two teams are winning.
Then three. Then ten. Then a hundred.

You didn't just build something that works.
You built something that could **be multiplied.**

This isn't theory. This is now.
This is what AI makes possible when we use it to **capture, organize, and simplify** what we've already built in obedience to God.

And the result?
A stronger Body.
A better-equipped network.
A collection of tools that any new team can pick up and run with—no matter where they are, how big they are, or how experienced they are.

Final Word: Don't Just Build—Document

If you want to build something that matters—**make it last.**
If you want to bless more people—**make it shareable.**
If you want to leave a legacy—**make it duplicatable.**

Your Mutual Success Team is part of something bigger.
And what works for you can work for others—if you'll stop and write it down.

Let AI help:

- Organize your thoughts

- Format your wins
- Break complex things into simple steps
- Create starter documents for new teams
- Store it in a Google Doc
- Share it with your network
- Submit it to TeamSuccessNetwork.com

God gave you wisdom. Let it become a well for others.

Because when we build together—and **document while we build**—the Church stops reinventing the wheel and starts multiplying impact.

Let's create documents that duplicate.
Let's make our progress easy to repeat.
And let's build a future where Mutual Success isn't just for us—it's for the **next team, the next church, and the next generation.**

Chapter 47 – Workflows That Multiply: Standardizing "Team Success" Processes with AI Tools

How to Build Repeatable Systems That Scale—So Your Mutual Success Projects Can Multiply Across Churches and Teams with Ease

---

One Process Done Well Can Change Everything

You've probably heard the phrase, *"Don't just work harder—work smarter."*
But what does that really mean for a church team, a ministry, or a Mutual Success Team trying to serve well, stay on track, and build something bigger than themselves?

It means this:
You don't need to invent everything from scratch every time.
You just need a **solid, repeatable process**—a *workflow*—that works.

Once that workflow is created, you don't need to *remember everything*, or train people from the ground up each time.
You have a **system** that can be duplicated.

This chapter will show you how to use **AI tools to build, refine, and standardize workflows** that help your Mutual Success Team:

- Work faster
- Stay aligned
- Get more done
- Onboard new members
- Share proven processes with other teams
- Multiply successful projects anywhere

Because in the end, it's not about working harder—it's about **working well together with wisdom.**

What Is a Workflow—and Why Does It Matter?

A workflow is simply this:

A step-by-step system that tells you what to do, in what order, and how to do it—so you get consistent results.

It's like a recipe for success.

Instead of chaos and confusion, workflows give you:

- Clarity
- Structure
- Focus
- Predictability
- Confidence
- Duplication

When a Mutual Success Team lacks clear workflows, you'll hear:

- "What are we supposed to do next?"
- "Didn't we already try that?"
- "I thought you were handling that."
- "We forgot again."
- "Let's just wing it."

But when workflows are in place, teams say:

- "Here's the checklist we follow every time."
- "That's already documented in our guide."
- "AI's tracking that in our weekly system."
- "Let's duplicate what worked in the last outreach."
- "I just sent the new member our onboarding doc—it's all there."

And that's where your team moves from **surviving to multiplying.**

How to Build a Workflow Using AI Tools

Creating your first workflows is easier than you think—especially when you let AI do the heavy lifting.

Here's a simple 5-step process to build a solid workflow using any AI platform:

1. Name the Project or Area

"We're creating a workflow for launching a fundraising project."

2. List the Main Phases

"Planning, Announcing, Launching, Following Up, Reporting"

3. Ask AI to Break It Down

Prompt:

“Help me create a workflow for a church fundraising project. Include 5 phases with 3–5 steps in each, and label who does each step.”

4. Ask for Formatting

Prompt:

“Now format that into a clear checklist that we can use again and again. Add time estimates and AI tools that can support each task.”

5. Save It in Google Docs

Title it clearly:

“Team Fundraising Workflow – Repeatable Steps for Future Projects”

Bonus: Create a **Template Version** that’s easy to customize for future teams.

Now you’ve got a full repeatable process that your team can:

- Open instantly
- Assign roles
- Automate with AI tools
- Duplicate for other projects
- Improve each time you use it

Examples of High-Impact Workflows You Can Standardize

Here are some workflows you can create using AI that will make your Mutual Success Team radically more efficient and duplicatable:

- **Outreach Workflow**

Invite, prep, post, pray, follow-up, share testimonies

- **Training New Members Workflow**

Welcome email, AI intro prompt, starter project, 15-min call, shared doc

- **Healing Testimony Workflow**

Capture story, ask AI to format it, add scripture, save in archive, share on socials

- **Bible Study Night Workflow**

Topic prompt to AI, flyer generation, 3-day reminder post, study guide, debrief

- **Mini Business Launch Workflow**

Brainstorm 3 ideas with AI, run SWOT analysis, plan 30-day test, write plan doc, share vision

- **Weekly Meeting Workflow**

AI-generated agenda → 20-min team check-in → assign action items → recap post → save doc

- **Prayer Request Workflow**

Collect forms → AI groups by topic → send to prayer team → log outcomes → praise report post

Every one of these can be built once... and then used **again and again.**

Final Word: Workflows That Multiply the Kingdom

When you build workflows, you're not just improving your team—you're preparing to **equip the Body.**

You're creating systems that allow:

- **New leaders to grow faster**
- **Smaller churches to operate stronger**
- **Your own team to stop re-inventing everything**
- **Your breakthrough to become someone else's foundation**

The power of workflows isn't just in the steps—it's in the *scalability.*

With AI helping you write, refine, organize, and store your workflows, you'll spend less time managing confusion—and more time building **real progress, real impact, and real Kingdom results.**

So, go build one.
Pick a process.

Ask AI to help you map it out.
Put it in a document.
Name it.
Share it.
Refine it.

You're not just leading a team.
You're creating a movement—**one workflow at a time.**

Chapter 48 – Now You Can Use AI to Its MAX!

You've Made It—Now Walk Boldly Into the Future with Faith, Vision, and Maximum Impact Through Your "Mutual Success Team"

You've Reached the Finish Line... But It's Really Just the Beginning

Congratulations. If you're reading this, you've completed an entire journey into one of the most exciting and transformative frontiers available to Christians today: the faithful, strategic, Spirit-led use of **AI to radically accelerate the Kingdom of Heaven coming to the Earth.**

You've seen what's possible.
You've discovered how to prompt with purpose, build workflows that multiply, organize ideas into action, document what works, and equip others.

You've learned how to empower your Mutual Success Team with clarity, creativity, communication, and compassion. And you've seen how AI can be more than a tool—it can be a **trusted partner**, a servant to your team's vision, and a magnifier of what God is already doing.

But now comes the best part:
You get to use it to the MAX.

What does that mean? It means this:
Don't stop at a few prompts.
Don't limit yourself to one idea.
Don't put a ceiling where God is saying,

"There's more. So much more."

Get Your Miracles

Get miracles out of it.

Get the deep, life changing transformations that are possible. Get AI to radically transform needed areas of your life. Don't stop with better. Go so much deeper, into the "real" improvements that COULD happen and that you really NEED.

Keep going and digging deeper. AI will only help you as much as you bend it to your inquisitive will. You can potentially get several life changing miracles with this revolutionary tool.

Then, you will be walking in a higher level of blessings that God wants for you.

Get a:

- Financial Miracle
- Relationship Miracle

- Better Relationships With Friends Miracle
- Better Business Relationships Miracle
- Learning New Things Miracle
- Becoming an Expert in Advanced Things Miracle
- Let it connect you to "real" miracles, finding Christian Resources – based on advanced AI Search features. It helps you to tap into the vast and rich resources around you all the time – near you, far from you, and everywhere. The internet is now more searchable than ever before.

Never stop.
Don't give up.
Seek the deeper answers that are always there, and always available.

There's No Limit to What AI Can Do—When Guided by the Spirit

People sometimes ask, "Is there a limit to how much AI can help?"
And the honest answer is: **Not really.**

Because AI is not a fixed tool. It evolves. It adapts.
And more importantly—**you evolve.** Your creativity expands. Your vision grows. Your prompting gets sharper. Your purpose becomes clearer. The Spirit leads you deeper.

So you're never stuck. Never capped. Never finished.

And best of all, **God isn't done either.**

He's still speaking.
Still guiding.

Still breathing life into your Mutual Success Projects.
Still moving through believers who are willing to ask:

"What's next, Lord?"
"How can we use this tool to serve You better?"
"What's the next breakthrough You want to bring through our team?"

When your heart stays yielded to God, your prompting becomes **prophetic.**
And your AI outputs begin to mirror not just your knowledge—but His wisdom.

We're Not Just Building Projects—We're Building a New Way of Living

The future isn't just digital. It's **relational.**
It's spiritual.
It's practical.
It's local and global—at the same time.

And through the guidance of the Holy Spirit, and the leverage of AI, we can now build:

- Businesses that fund missions and meet real needs
- Projects that multiply across churches
- Training that works in every town, village, and nation
- Curriculums that teach truth and transfer wisdom
- Content that inspires, uplifts, and transforms lives
- Support systems that ensure **no Christian is left behind**

We're not just launching ideas.
We're launching **a movement**—the Mutual Success Movement.
Where churches don't compete—they complete.
Where believers collaborate.
Where solutions are shared.
Where the wins of one become the wins of many.

And every one of us has a part to play.

The Kingdom Is Near—Let's Build It Together

Jesus said, *"The Kingdom of Heaven is near."*
He wasn't just talking about spiritual salvation—He was talking about **the reality of Heaven touching earth.**

Where people are healed.
Where needs are met.
Where justice rolls like a river.
Where abundance flows.
Where knowledge is multiplied.
Where unity thrives.

That's the world we're building.
That's the Church we're becoming.
And that's the work AI helps us accelerate—if we'll use it **to its max**.

Here's how you can step into that starting today:

- Use AI daily—not just as a helper, but as a partner
- Share what you're doing with other churches
- Turn your success into documents that can be reused

- Send your breakthroughs to the Team Success Network
- Help eliminate global need—one step, one person, one idea at a time
- Inquire of the Lord for every decision—and follow with action
- Stay humble. Stay bold. Stay connected.

And remember:
You are not alone.

There's a whole movement of believers just like you—faithful, curious, obedient, and ready to build the future.

Together, we are a **Global Mutual Success Team.**

Let's move forward. Let's build together. Let's multiply the miracles.
Let's bring Heaven closer, one decision at a time.

You're part of something bigger now.
You've been equipped.
You've been commissioned.

Now go use AI to the max—**and bless the world.**

Want to get connected? Email us at **connect@TeamSuccessNetwork.com**.

Want to share "real" results - of how AI helped you in a major way, as a result of this book? Email us at **results@TeamSuccessNetwork.com** and be part of the next chapter.

Books By the *Team Success Network* - In The "Team Success" Series:

Book 1 - Kingdom of Heaven – Team Success Network

Book 2 - Building Duplicatable "Mutual Success Team" Business Projects

Book 3 - Team Success Network: Leadership, Development, Summits, & Discipleship

Book 4 - AI Tools - Team Success Network - Enhancing Mutual Success Teams Using AI Tools

Book 5 - Team Success Network: Global Organization & Resources

Book 6 - 200+ Business Project Ideas – For Mutual Success Teams

Book 7 - The FAITH HEALING TRAININGS - Learn Healing As a Skill

Book 8 - Using Christian Decks as Your First "Mutual Success Project"

Meet the Author – Elijah Jonathon Stone

Founder of the Team Success Network

Hi, I'm Elijah. I wrote this book because I believe God is calling the Church into a new era—where we don't just gather, we build. I'm not a celebrity, a pastor of a mega church, or a spiritual expert. I'm someone like you—who saw the need and heard the call.

The Team Success Network is not just a concept. It's a movement of real believers working together to bring God's Kingdom into real life—through unity, purpose, and mutual success. I hope this book encourages and equips you to step into your role.

If this vision speaks to your heart, I'd love to hear from you.

Email me personally: elijah@teamsuccessnetwork.com

Let's keep building—together.

Stay Connected – Your Next Step Starts Now

You've made it through this book—but that doesn't mean the journey is over. In fact, it may just be beginning.

The Team Success Network was created for people like you—believers who are ready to live differently, work together, and take action that transforms lives. If something in these pages sparked a new idea, birthed a desire to join or lead a team, or helped you experience a breakthrough—we want to know. Your story matters.

Let's move from inspiration to action.

Reach out to us:

📩 To Connect – connect@teamsuccessnetwork.com

📩 Share What God Is Doing – results@teamsuccessnetwork.com

📩 Send Us Documents or Team Reports – docs@teamsuccessnetwork.com

📩 Write Elijah Personally – elijah@teamsuccessnetwork.com

You're not alone in this. The Kingdom is advancing—and we're building it together. Let's keep going.

Made in United States
Troutdale, OR
11/03/2026

56844704R00176